B

Reinhard Breuer

The
Anthropic
Principle

Man
as the Focal Point
of Nature

Translated by
Harry Newman and Mark Lowery

Springer Science+Business Media, LLC

Reinhard Breuer
Daimler Benz AG
FT/P
Postfach 80 02 30
D-7000 Stuttgart 80
Federal Republic of Germany

Library of Congress Cataloging-in-Publication Data
Breuer, Reinhard A., 1946–
   [Anthropische Prinzip. English]
   The anthropic principle : man as the focal point of nature /
[Reinhard Breuer].
      p. cm.
   Translation of: Das Anthropische Prinzip.
   Includes bibliographical references (p.
   ISBN 978-0-8176-3482-7
   1. Cosmology—Popular works.   2. Physics—Popular works.
I. Title.
QB982.B7413   1990
523.1—dc20   90-1121
                                                 CIP

Printed on acid-free paper.

Originally published in german as Das Anthropische Prinzip: Der Mensch im Fadenkreuz der
Naturgesetze, by Reinhard Breuer. Copyright Meyster Verlag Gmbh, Wien – München, 1981.

© Springer Science+Business Media New York 1991
Originally published by Birkhäuser Boston, Inc. in 1991

9 8 7 6 5 4 3 2 1

ISBN 978-0-8176-3482-7   ISBN 978-1-4899-6741-1 (eBook)
DOI 10.1007/978-1-4899-6741-1

For Willi Ochs

"We are here in this quite fantastic universe
and have hardly an inkling of whether our existence
has a real meaning."

Fred Hoyle

"As we look out into the universe and identify
the many accidents of physics and astronomy
that have worked to our benefit,
it almost seems
as if the universe must in some sense have known
that we were coming."

Freeman J. Dyson

Foreword

Our minds are unwilling to contemplate a second or third universe besides our own. In some languages, the words for "universe" and "cosmos" will not take a plural. This is probably because there is no point in talking of other universes alongside our own unique cosmos — universes which we can know nothing of, since everything we observe must by definition be part of the universe we inhabit. We can talk of other planets since we can perceive them from here; we can speak of other suns and other solar systems, other forms of life than those familiar to us. It is only in discussing other universes that we get into difficulties. These no doubt arise from the fact that our concept of the universe is too dependent on our three-dimensional notions of space.

Leibnitz did not have this problem. "We live in the most perfect of all worlds," he said, and the concept of alternative worlds came easily to him. For here our world is compared with other *conceivable* worlds, and not with ones supposedly realized elsewhere. His worlds are conceptual models to be thought of as existing *in place of* our own rather than outside of it. They contain natural laws and life appropriate to themselves: laws which need not necessarily be the same as ours. It is possible to conceive of a universe in which there is no electrical force, or no gravitation. After considering all thinkable possibilities, the real universe around us struck Leibnitz as the most perfect (although we may suppose that a universe without an Isaac Newton, from whom he continually suffered direct and indirect attacks as a result of the battle for priority of invention of the Differential Calculus, might have appeared in some degree even more perfect). The astronomer Johann Heinrich Lambert (1728-1777), a contemporary of Kant, raised this maxim to an astronomical article of faith. His most perfect world is the one in

vii

which the most people can live, and so glorify creation. The law of gravity must be such that planets cannot collide, for then admirers of creation would be destroyed. In Lambert's universe God has so arranged the laws of nature that humanity and all sentient inhabitants of other planets derive a maximum of health and happiness. In this, he has already arrived at something like the Strong Anthropic Principle. The laws of nature are deliberately set up in such a way that life is preserved. Today it is formulated in a different way: The laws of nature applying to our universe must be such that life can arise under them and endure for long periods of time, for if it were otherwise we would not be here.

It is this principle that the following pages address. It does in fact appear as if the laws of nature were set up in such a way that life was bound to emerge and life could develop. A change in the numerical values of the universal constants would result in a different universe which would be unlikely to support life. It seems as if a large number of accidents had to conjoin to create the conditions for our existence. Breuer lists these, and it is an impressive list: we have the feeling that this number of accidents could not happen by accident. It is almost as if the laws of nature were set up with us in mind.

Needless to say, this idea must be a difficult pill for any Copernican to swallow. My own doubts urge me to make the following comparison. Yesterday I met a friend of mine. He was wearing a red tie. I know that he possesses twelve ties; the probability that he would be wearing the red one is thus one-twelfth. He had on his feet his leather brogues; as he has eight pairs of shoes this give a further probability of one-eigth. The probability that he would be wearing this tie *and* these shoes is $1/12 \times 1/8$, that is 1 in 96. He was wearing a certain one of his 24 shirts... When I had finished all my calculations it was clear that yesterday's meeting was so unlikely that I began to doubt whether I had in fact met him at all. Isn't this exactly how it is with life in our universe? Once we are here, we begin to discover how improbable it is that we should ever have come into being. But here we are.

Although it appears that nature is so arranged as to favor life, we must beware of falling prey to the logic of the medieval monk who averred that we should be grateful to God for arranging things so that the sun shines during the day rather than at night when it is no use to us. Is it not perhaps possible that life could have adapted to suit any universe, just as Man has adapted to the alternation of night and day, so that maybe there is no reason to be surprised at our existence at all? Perhaps we would have achieved an existence in any universe — even one with the craziest of universal constants.

Breuer explains the conditions necessary for life and shows that universes with alternative physics could hold little hope for life. But Breuer's "alternative worlds" have come about only by assigning a handful of natural constants different numerical values, while the physical laws themselves maintain their structure. Worlds are also conceivable in which the natural laws have absolutely no similarities with ours (and none with the laws of nature which remain if one omits two or three of the four fundamental forces.) Concerning what is valid or not in such a world, we have as yet spent no time exploring that boundless ocean of possibilities.

The anthropic principle gives us the stimulus to reflect on ourselves, the laws of nature, the cosmos. Breuer's book leads us to the beginning of the universe, when matter itself came into existence, to galaxies and their stars in which the atoms essential to life were cooked together, and to the planets, on which complex biological molecules could be formed. Overall it appears as if from the beginning Nature was directed towards us, as if it was working toward our arrival. Are we succumbing to a fallacy here or are we on the trail of a not yet understood new truth?

A variation of a passage from Brecht's *Galileo* comes to mind: Thoughts of the anthropic principle have been pestering me like an itch ever since I first read about it.

Rudolf Kippenhahn*

---

* Rudolf Kippenhahn is Director of the Institute for Astrophysics at the Max Planck Institut for Physics and Astrophysics and an honorary professor at Ludwig-Maximilians-University, Munich.

Note by the Author

When it appeared in the German original, this book was the first work devoted entirely to the anthropic principle. In the interim between then and publication of this English version, much has been learned. I have updated the material where practical and given reference to books on the subject which have appeared since.

Hamburg, June 1990
Reinhard Breuer

Contents

Chapter I
Man and His Place in the Universe

MAN THE MEASURE OF ALL THINGS?

How must the universe be constructed in order to produce life? Can we imagine a different cosmos in which intelligent life might also appear? What preconditions were the deciding factors for the origin of life? The one thing that makes our universe special is the fact that it has produced intelligent observers. But, looking at it the other way, only an intelligent life form can study the universe and probe the causes of its origin.

What conditions had to be fulfilled by the cosmos and the laws of nature to bring forth a life form able to recognize these conditions? On Earth, it is Man that has made the universe "self-aware," or "self-cognizant." This self-awareness has shown that microcosm and macrocosm must have cooperated via a wealth of "accidental" interrelationships to make terrestrial life possible. The same applies to alternative and artificial life forms. The laws of nature, the origin of matter, cosmic expansion and biological evolution appear to have functioned as a team with such fine precision, demonstrated in such a variety of ways, that intelligent life could only come about through the operation of the entire "machinery of the universe."

This intimate bond between natural law, the cosmos, and the existence of life on Earth — the only example of intelligent life that we know of in the universe — has become known as the "anthropic principle." By applying this principle to the physical world, we can pick out from among the various hypothetically conceivable universes the

1

particular one in which life is possible. The result is the hypothesis presented in this book: life — as we know it — is only possible in the universe we know!

This hypothesis can be illustrated by the following question and answer:

"Why is the universe as old as it is — ten to twenty billion years?"

"It is that old because we couldn't be here much before now."

This kind of example is an integral part of the interrogation of the universe in recent years being embarked upon by physicists, one which consistently takes as its basic starting point the fact that we are here. The existence of the human race limits cosmos and nature in their ability to be "different" from what we actually observe. So is Man "the measure of all things?" Are we the yardstick by which we should assess all natural phenomena, even such basic aspects as the fundamental laws of the universe to which everything in it owes its being and its form?

This book aims to explore how far one can get in analyzing the structure of the universe by application of the anthropic principle. It will make clear that this principle is — despite its somewhat parodoxical appearance — no tautology; nor can it be employed for a teleological interpretation of the world. Instead, the anthropic principle is a legitimate tool of science with a potential for prediction. In more general terms, we will have to explore the unique and many-threaded relationships between Man and his surroundings on the one hand, and the wider structure of the cosmos on the other — where we mean by his *surroundings* the Earth, the solar system, our galaxy and the cosmos, their influence on one another and their evolution, and by *the structure of the cosmos*, the laws of nature and the strengths of the fundamental forces. In this context, the question that has been asked is: what forms can life take in this universe? However, following an idea by the Princeton physicist Robert H. Dicke, I shall turn this question back to front and attempt a preliminary answer, starting from the other end, so to speak:

On the Earth there is a life form with consciousness, an observing intelligence. What form must necessarily be taken by the universe in which it lives? This question cannot be answered without following these logical steps:

2

— Consciousness should be tied to some form of life;

— Life requires chemical elements, especially those heavier than hydrogen and helium, before it can come into being;

— But these heavier elements are only produced by thermonuclear combustion of the lighter elements — that is, by the fusion of light atomic nuclei;

— However, nuclear fusion only takes place in the interiors of stars and requires a minimum of one billion years to produce appreciable quantities of heavy elements;

— But a period of a few billion years is only available in a universe endowed with a gravitational force weak enough to permit eternal expansion. In the event that such a universe re-collapses, it should be at least this large at the moment of its greatest extension;

— At later stages in the life of the cosmos, on the other hand, sun-like stars would be extremely rare, most stars being low-energy white dwarfs which would be unable to supply sufficient energy for a planetary life form undergoing slow biological evolution.

Thus, an answer to the question why the universe we observe today is just as old as it is could be: because if it were not, life would not be here.

This example stands for many others involved in the discussion. It is not too much of an exaggeration to turn around the statement that human life only exists because the universe and the physical world have certain special characteristics and arrive at the following claim: the universe and the physical world must, taken as a whole, be constructed in a certain special way to be able to produce at least one type of "intelligent" life form.

The seemingly paradoxical and mostly unorthodox argumentation of this "anthropic principle" found a place in the world of science in the '70s. The new approach seeks to draw scientific benefit from the fact and the characteristics of human existence.

The reversal of the usual way of looking at things that this involves — at least as far as the origin of life is concerned — is not without a certain boldness; indeed, if taken to its logical extreme it would bring

3

about a historic turning point in the evaluation of humanity's place in the cosmos. It sets right the relationship between the observer and the world he observes and throws light on the unity between nature and the observers it has produced.

The anthropic principle explicitly utilizes characteristics of terrestrial life, and in the process takes as its sole starting point an observation which, though obvious enough to us, is particularly wide in its consequences. But, we might ask at this juncture, isn't this a retrograde step for human culture? Doesn't it run contrary to the developments of the three centuries since Nicholas Copernicus expelled the Earth from its throne in the center of Creation? Haven't we withdrawn from our anthropomorphic presumptions and self-aggrandizement to a more modest place in the cosmos, one of many civilizations, on a planet like many others? What is so special about life on Earth that can even partially justify a reversing of the Copernican revolution? In the universe aren't there any number of other life forms, differing in chemistry, using elements other than carbon, nitrogen, and oxygen as a basis and provided with a different genetic code?

These objections appear weighty. But it requires only a little thought to see that the existence of other, totally different, conscious life forms would limit the freedom of the universe to be different from the way it is, not to a lesser degree, but to a *far greater* extent than the existence of the *human race alone* already does. If life could arise in the universe under widely differing circumstances and find a variety of routes to intelligence, and especially if life were possible under totally different natural laws, then Man would no longer be the sole "measure of all things." There would then almost certainly be further cosmic "accidents" that would need to have come about before these other life forms could have arisen — preconditions which would of necessity overlap and interact with the "accidents" necessary to the origin of terrestrial life. In such a case — as will be explained later — the anthropic principle would be in an even stronger position.

The facts of the case, however, are that although we may not occupy a central position, we are in what is in many respects a special one. It can be argued that the manner in which terrestrial life is constructed gives it a significant place in the range of possible types of life. Take, for instance, the most important chemical elements from which life on Earth is built. Terrestrial life can be characterized in chemical terms by the properties of the carbon atom and water as a chemical solvent. A variety of reasons suggests that hardly any other of the one hundred or so chemical elements is so well suited to the construction of a biology as complex as that found on Earth. This is a major obstacle to speculations concerning life forms based on silicon (crystals) or deriving their energy from breathing methane, even though this latter alternative to our present-day chemistry of life did indeed exist on Earth as late as a few billion years ago. At that time, when the atmosphere was dominated not by oxygen but most probably by carbon dioxide, methane and ammonia, it is likely that methane-breathing organisms were widespread; but this branch of evolution, like others whose final offshoots we still find preserved on the level of bacteria in a few ecological niches, never progressed beyond the stage of single-cell bacteria.

Considerations such as this point to carbon as the basic element and water as the basic solvent for biological systems in our universe. (More on this subject can be found in my book on the problem of extraterrestrial life, *Contact with the Stars*). Another essential ingredient for the development of higher life forms is a storage-and-retrieval system ("memory") for biological information. In terrestrial biology this function is carried out by the biomolecules of the celebrated DNA, which store biological information on inherited characteristics in reproducible macromolecules taking the form of a twin, intertwined spiral — the famous "double helix." Within the constraints of the fundamental forces we know, which control the chemical bonding and spatial arrangement of these molecules, it is difficult at present to imagine any alternative basis which would allow the development of a memory that would match their efficiency. (See chapter VIII).

Of course, in saying this we may simply be showing our stupidity and ignorance. It is always open to us to postulate intelligent life forms of a totally different nature from ourselves — both under the

5

laws of our universe and in a hypothetical different universe ruled by completely different natural laws. In principle, of course, the scientist must always keep an open mind on hypothetical alternatives; but they can only be of limited interest to us as long as we are unable to offer any precise statements regarding these alternative types of life. It is another matter with the genetic code, the "grammar" of reproduction and chemical interaction between macromolecules, which enables the information stored in DNA to be employed in the process of inheritance. Manfred Eigen of the Max Planck Institute for Biophysical Chemistry in Göttingen, West Germany, considers that the genetic code arose statistically on the young Earth in the self-organization of a large group of molecules into a cooperative system. That is to say, it is possible that a number of these cooperatives came into being on our planet independently of each other, spread out in their respective areas and eventually came into competition with one another, until one of them prevailed and became the sole survivor, *the* genetic code of terrestrial life. Remnants of a different genetic code were discovered in a bacterium in 1980.

A different genetic mechanism of inheritance, then, is quite conceivable. If life could start up over again on Earth, it would probably possess a different genetic code. But this would have little effect on the principle of biochemical information storage by macromolecules, since large molecules of this type, with a structure analogous to that of DNA and RNA, could still have been built up. Comparable and equally efficient biological memory systems would have to have come about in order to store and pass on to the next generation batches of information of as many as several billion bits (as in human genetic material). Thus, the more the bandwidth of possible structures of life is limited and the more its boundaries are marked out by chemical and biological considerations, the more terrestrial life becomes the sole factor limiting the universe in its possible structure. This also applies to any alternative biology not too different from our own. Unfortunately, it is only the life forms of Earth that we know anything at all about, and only in regard to relatively minor variations from this can we make statements which are even halfway meaningful in scientific terms. Beyond this —

in the case of a hypothetical and totally different life form — we would be venturing into a twilight of untestable, uncriticizable, and thus, to science, "unthinkable" speculation.

The questions below, in their different aspects, give us in brief the subject of the following discussion:

— When do the laws of nature permit the development of intelligent life?
— When would the laws of nature on no account permit the evolution of life?
— Even given that all the laws of nature were from the beginning "at an optimum" for the development of life, under what conditions does evolution in fact begin, leading from the molecule to intelligence?

Among other things, we shall consider how celestial and terrestrial catastrophes have advanced, and frequently initiated, an evolutionary process — as, for instance, with the origin of the elements in the Big Bang and in the explosions of stars and the formation of galaxies, stars, and planetary systems — along with the relationship between the four fundamental forces of nature, elementary particles and the biochemical foundations of a life form based on carbon.

The conclusions we reach are: if it is required of the cosmos that it shall produce intelligent life, then the laws of nature, the structure of the universe and the structure of human life are almost irrevocably fixed. Our cosmos has essentially the characteristics we have observed around us in nature. Every universe differing from this would remain uninhabited and could never attain the state of cosmic "self-awareness." "And as we look out into the universe," says Freeman Dyson, "and identify the many accidents of physics and astronomy that have worked together to our benefit, it almost seems as if the universe must in some sense have known that we were coming."

THE STRONG VS. WEAK ANTHROPIC PRINCIPLE

Is it only by accident that life exists in our universe? Or are there considerations which would make this phenomenon seem less probable? Undeniably, we are here, observing the universe. Taking this as our starting point, we can begin by looking for the conditions the universe would need to fulfill in order that we could be here to observe it.

To answer this question in a methodical way, it will help to reformulate it as a principle, that is, as a requirement — or working hypothesis — open to refutation. We can start with the following statement: "what we observe must be limited by conditions necessary to our presence as observers." A universe that provides for the existence of an intelligence which then observes it deserves a special name: it is a *self-cognizant universe.* The statement just quoted may, in retrospect, seem trivial; but if we look at it in another way it will serve to open our eyes and enable us to look at the world around us from a new basic standpoint; and exactly because it is trivial one falls into error if one does not take it into account. It will be clear that our existence is inseparably bound up with our environment, the variety of living things, the shape of our planet, the astronomical class of stars to which the sun belongs, the cosmos and the fundamental forces of physics.

But how mutually dependent is this relationship? Are the factors involved so completely interwoven that not a single property of the cosmos could be changed without reducing the existence of mankind to a vain hope? Or are there a few basic structures that are essential to the existence of life and intelligence? The closer the relationship between the human race and its cosmic environment, the more inevitably the properties of the natural world and the cosmos must derive from our study of ourselves. Formulating this as a "principle," we can begin by stating:

Weak Anthropic Principle
Because there are observers in our universe, the universe must possess properties which permit the existence of these observers.

In other words, the universe must be consistent with the fact that it contains observers (whose position will necessarily be privileged).

Strong Anthropic Principle
The structure of the universe and the particulars of its construction are essentially fixed by the condition that at some point it inevitably produces an observer.

8

The weak version of the anthropic principle was suggested by Robert H. Dicke in 1961. It is the one we shall consider throughout this book unless otherwise stated. The conceivable extension of this (as formulated in the strong anthropic principle) was not examined by Dicke; from his standpoint, then, the universe did not necessarily have to produce an observer. But as this has unquestionably come about, we can infer a series of conclusions from it and deduce the conditions necessary to our existence.

The second of these hypotheses owes its origin to the British physicist Brandon Carter of the Observatory at Meudon, outside Paris, who presented it at a meeting of astronomers in Krakow, Poland, in 1973. In investigating the interrelationships between observer and universe, either principle can be applied in at least two ways. Firstly, by changing the terrestrial conditions under which life began and following the evolutionary process through in this new context; secondly, by altering not only planetary conditions, but also introducing hypothetical changes into the cosmic surroundings. In addition, the strong principle allows for the most fundamental intervention by altering the laws of nature and the fundamental forces of physics, which are then, so to speak, implemented in a different universe and its hypothetical development analyzed.

What scientists discovered when they did this was that the most important properties of galaxies, stars, and planets, as well as those of our immediate surroundings, are essentially decided by the fundamental forces of the microcosm, along with the effects of gravity. Of course, many relationships between quantities in different branches of science, although at first surprising, can be shown to be the direct result of simple physical considerations. But other aspects of our universe, some of which seem indispensable to the evolution of any form of life, depend in a complex way on relationships between the "physical constants" which formerly seemed accidental — the now familiar "coincidences" (see Chapter II).

Among these constants are the speed of light, the gravitational constant, the elementary charge on the electron and the masses of the most common atomic particles. These startling cross-connections between apparently unrelated fields of science are made at least in part more explicable by applying the anthropic principle — indeed, scientists ought to have been able, with its help, to "predict" them. As Brandon Carter remarks: "...these coincidences should rather be considered as confirming conventional [general relativistic Big Bang] physics and astronomy, which could in principle have been used to predict them all in advance of their observation."

This "prediction" of apparent coincidences, however, is only possible by applying the anthropic principle. If the principle is used in a scientific study, three types of theoretical prediction are possible:
— Predictions relying on conventional physics and chemistry, without the support of the anthropic principle;
— Predictions in which the weak anthropic principle is used;
— Predictions using the strong anthropic principle. By keeping all three options open, the researcher avoids the temptation to apply the anthropic principle indiscriminately to account for anything "inexplicable" or "coincidental," but is encouraged to look first for explanations based on current physical theories. The following pages will demonstrate just how many astonishing characteristics of the physical world are no more than the expression of basic physical laws. Not everything we see around us is exactly the way it is simply because we exist and are observing it. Rather, the anthropic principle sets us at the beginning of a wider appreciation of nature, not at the end; and it could well happen that some phenomena that only seem explicable and inevitable to us at present if we invoke the anthropic principle may one day be explained in a quite traditional manner, as the consequence of a physical theory.

Why is it that scientists are already set on adding a new principle to the edifice of traditional physics, when it has hardly begun to resolve itself from the fog of unexplored ideas? Quite simply because it has become apparent that many properties of the physical world can only be understood if use is made of the fact of the existence of intelligent primates, specifically man, as a biological-geological datum. Sounding out the structure that unites man and nature to learn more of its internal foundations and buttressing is one of the most exciting adventures in modern science. For the first time, the anthropic principle allows us to examine the inner consistency of science's picture of the universe and discover the close interrelationships between three spheres — the fundamental laws of nature, the conditions and evolution particular to this cosmos, and the existence of an intelligent observer.

MAN'S PLACE IN THE UNIVERSE

The anthropic principle does not place the human race in any (spatially) central position with respect to physics and cosmology, but it does give us a special role, since our existence, at least on this planet, is without question something special. Civilizations have existed on Earth only in the last few thousand years — a brief moment in cosmic terms, compared with the age of the earth: five billion years. And humanity only developed a *technological* civilization one hundred years ago. As a civilization on this level, we are a rare or even unique phenomenon, at least in our immediate cosmic neighborhood, which stretches some hundreds or thousands of light years around us. Intelligent but non-technological civilizations may be more common; but we cannot observe them since they cannot make themselves noticed, whether by signals or by indirectly observable technological activities.

If the role of the human race is drawn into scientific observations of the basic physical elements of nature, we shall be returning in some degree to a long-abandoned position, the cosmology of the Greek philosopher Ptolemy. Taking his lead from even older concepts, he had set the Earth in the center of the universe, although by this time Aristarchos of Samos had already discussed a model of the universe with the sun as its pivot. With Nicholas Copernicus, who took up Aristarchos' idea, the Earth, and humanity with it, was banished from the center of cosmic events — a heavy blow to the self-esteem of his contemporaries. Suddenly, the sun was to be the center, and all the planets, the Earth and the stars were to move round it.

A number of modern-day cosmologists went even beyond Copernicus. They insisted that the Earth should take an indistinguished place in the cosmos not only in space but in time as well. The present cosmic epoch, they proposed, was in essence no different from past or future chapters in the development of the universe. This demand would support a cosmos with no Big Bang and no true cosmic evolution, but only eternity and immutability — a model that was accepted as late as the 1920s. The discovery of galactic regression (cosmic expansion) by the astronomer Hubble in the 1920s and that of cosmic background radiation in 1965 by Penzias and Wilson (see Chapter IV) pulled the ground from under such continuing attempts at "humiliation" to some extent. With the advent of the anthropic principle, a certain revision of this standpoint is under way. As far as possible, it finally gives credit to the special role in the cosmos played by Man — through the point in time at which we exist; through the place of our existence in a particular planetary, solar, and galactic neighborhood; and beyond this, through our evident close relationship with natural events, whose varied and subtle interaction was necessary before our existence could come about — Man as the linchpin of nature.

THE ANTHROPIC PRINCIPLE — OPTIMISM VS. PESSIMISM

The anthropic principle can be approached from at least two standpoints — an optimistic and a pessimistic one. From the optimistic point of view, life is such a frequent phenomenon, with such a high probability and such adaptability that some form of life equipped with intelligence will practically always develop. Also implicit in this is the view that we are not in fact capable of imagining life forms differing from ourselves — whether arising under the same conditions or different ones. For this reason our argumentation will necessarily always be "anthropomorphic": it will always be dazzled and biased by this unique terrestrial example.

In opposition to this is the pessimistic standpoint, as represented by George R. F. Ellis of the University of Cape Town. "The existence of life in general," says Ellis, "and intelligent life in particular, is an incredibly unlikely eventuality, both in terms of the possibility of its existence (that is, the compatibility of the possible structures of intelligent life with the local laws of physics) and of the probability of its evolution."

This viewpoint also asserts that "there will be bounds on the variation of each of the constants such that if these bounds are exceeded, life will not be possible at all." And if we consider only variations within these bounds, the possible types of life — and therefore the possible routes of evolution — would already be different from those possible without such variations. Optimism and pessimism: the fact that such diametrically opposing views can be taken on this subject simply demonstrates how little we know about alternative life forms and evolutionary routes. However, the following chapters aim to show that certain changes in the constants involved in the laws of nature are prejudicial to the conditions indispensable to the origin of life (that is, life as we know it). This, as we shall demonstrate, implies support for the following thesis:

It is unlikely that any form of life can come about if the fundamental constants have substantially different values from those prevailing in our own cosmos.

How Do We Change the Laws of Nature?

How can we make different laws of nature than those that apply? In what way can we deviate from the architecture of nature recognized to be correct, and still make some sense of the result? The substance of physics is expressed above all through its (mathematical) laws and through a set of constants that appear in these laws. A scientific theory will also be provided with rules by which its concepts are to be interpreted — mathematical equations and predictions. An important element of the anthropic principle concerns the dependence of an observer of the cosmos — in a word, our own existence — on the structure of the laws of nature. This relationship can be tested by varying the laws of nature and noting at what point this variation makes the existence of the human race impossible. If this test is to be done properly, it must be preceded by a pause for thought concerning just what sort of changes can be made in the laws of nature without making it completely impossible to discuss the resulting scenario in any sensible way; the initial flights of the imagination must be reined in.

13

Without question the easiest way to set about it is to vary only the numerical value of the natural constants without interfering with the *form* of the laws of nature. This will above all affect the strength of the individual natural forces. Of course, the form of the natural forces could equally be altered, but such a thorough tampering with their structure would make it increasingly difficult to calculate the effects on the resulting universes. In order to grasp at least the most important relationships, it is best to set up the most simple program that can be designed, so long as it is practical to carry it out. This will leave the laws of nature in the form in which we currently consider them to be correct. The changes are made only in the fundamental constants, and principally in the strengths of the four basic forces of nature. Under the new conditions created by these changes, scientists can estimate what influence the changes will have on the processes leading to the origin of life and intelligence.

How do the fundamental constants come to be involved in the laws of nature in the first place? Usually their involvement takes the following form. In a mathematical equation, one formula is set equal to another, but linked by means of a "proportionality factor." A simple example is the formula which states that the distance traveled by light is proportional to the time it takes to cover the distance. The proportionality factor in this is a constant, in this case, the speed of light:

$$\text{(light distance)} = \text{(speed of light)} \times \text{(light time)}$$

More generally, and in a less trivial context, the fundamental constants most frequently appear in the following form:

$$\text{(mathematical formula)} = \text{(constant)} \times \text{(second formula)}.$$

Examples of this are the field equations in the theories of Newton, Maxwell, and Einstein. Newton's law of gravity, for instance, states the force with which two bodies a given distance apart will attract each other:

$$\text{(gravitational force)} =$$
$$\text{(gravitational constant)} \times \text{(mass1)} \times \text{(mass2)}/\text{(distance)}^2$$

14

Here the gravitational constant acts as a factor converting two masses and the distance between them into a force of attraction. This numerical factor, then, is the element that largely decides the strength of gravitational force. The larger the value, the more strongly the two bodies will attract each other, and the smaller it is, the weaker the attraction will be. There are, of course, other cases where the constants act within a more complicated relationship.

However, the simple form illustrated — and this is quite adequate for most of the basic laws of physics — provides the physicists obsessed with the anthropic principle a simple way of interfering in the laws of nature and thus in the basic structures of the cosmos. The mathematical content — that is, the formula — can be left untouched while the experimenter restricts himself to altering the constant, "adjusting" it like a screw. The constants are numbers, but provided with a dimension such as length (centimeters), mass (grams) or time (seconds). The presence of these dimensions should not mislead anyone into thinking the numerical values of the constants may be varied simply by, for example, changing the expression of the speed of light from centimeters per second to — say — yards per week!

Such an apparent variation of the constants is naturally not what we are thinking of. This would have nothing to do with physics, but would be no more than a change in the conventional way of expressing the value. This false lead can be avoided if the dimension is taken away by expressing the constants in dimensionless numbers. The simplest way of doing this is by observing relationships between quantities similar in kind: thus, the speed of light can be given in relation to another speed, or the mass of the electron as its relationship to another atomic mass. Any change in these pure numbers will have a real influence on physics and go far beyond the bounds of a mere cultural convention: it will affect the laws of nature in a quite specific, though relatively simple manner. The law will be preserved in its familiar form, but the influence of the second mathematical formula on the first will be different — according to the change that has been made, its force will be greater or smaller, or even, if the constant is assigned a zero value, removed altogether.

As the "clockwork of nature" is driven by four different forces, for which there are four corresponding interaction constants, there are, so to speak, four adjusting screws by which the interrelationships of the forces of nature can be regulated. (For the sake of simplicity, it is assumed that these four screws controlling the mainsprings of the universe can be adjusted, along with other constants, independently of one another, and are not coupled together in some unknown way, or perhaps after the manner of a wider, "unifying" theory). This procedure may be simple to look at, but can have far-reaching effects. As each changed constant will alter the strength of the corresponding natural force, the "adjustment" in question will affect the orbits of the planets, the origins of the chemical elements and the speed of biological reactions, along with evolutionary processes involved in the construction of cells and other organic systems. This kind of intervention in the natural forces, therefore, permits us to create for ourselves, as it were, a new universe, whose development from the Big Bang onwards (if there is one in the changed universe) can then — in principle, at least — be mapped out in calculations through to its end, either with pencil and paper or using the astrophysicist's favorite toy, a computer.

This is not an impossible task. Because the changes in the laws of nature have been carried out in a deliberately simple way, these alternative universes can be followed through relatively precisely and their most important properties calculated, down to the possibilities for the birth of an intelligent life form.

ANTHROPIC/ANTHROPOCENTRIC

(A CRITIQUE OF THE ANTHROPIC PRINCIPLE)

The insights hoped for from experiments of this type particularly concentrate on the place of life in our universe. Would it be possible for life, intelligent life above all, to come about in practically any given cosmos? Or does it demand a quite particular and special cosmos? Could it be that the human race exists simply because the universe happens to be the way it is? (And what do we mean when we say it "happens to be?") Isn't this, superficially, a trivial, even anthropomorphic-sounding result? ("Anthropomorphic" is the term used for ways of thinking that result from being too bound up with our own terrestrial origins and are as a consequence inevitably rather blinkered.)

The weak anthropic principle is innocent of such anthropomorphism, but the strong anthropic principle cannot entirely escape the accusation. Admittedly, it is difficult to study the fundamentals of our own existence — and it is an even more complicated matter to look beyond its framework and reflect on the nature of our nature. However imbued we may be, individually and as a race, with the capacity for reflection, to what extent are we at all able to give meaningful thought to an entire universe hypothetically constructed on different principles, and follow through its development? And more: how far can we escape the suspicion sown in us by the evolutionary theory of knowledge, that we only perceive the existing world the way we do because after millions of years of adaptation we have developed in parallel with it to such an extent that we carry its structures within ourselves — not only materially, but intellectually as well? It must remain for the present an open question whether we are cognitively capable of determining the properties of a universe which differs substantially from our own, but which could nevertheless produce a life form to observe it.

Our modest goal in this book is that of tracing back to the foundations of natural events the essential characteristics of one life form which have made it an observer of this universe. Beyond this, our task is to establish, at least in a rudimentary way, how the universe could *not* have been constructed if it was to become the stage for an intelligence comparable with ourselves. If, in the extreme case, the fixing of structure were unique and we could drop the word "essential" from the definition, then the strong principle could be rephrased: the universe is as it is because otherwise it would not produce observers. In terms of the weak principle, on the other hand, it can be shown how Man and life on Earth in general could be used as a kind of sensor for the properties of the universe. For many it seems that the weak anthropic principle appears rather trivial. Naturally, they say we are here because nature has produced us; and hence nature and the universe can not be in contradiction to our existence. Brandon Carter of the Observatoire de Paris at Meudon therefore prefers to call it a "Principle of Self-selection." "Exactly because it's trivial," he explains, "we have to take it seriously. If you forget about it you make mistakes. If you make use of it, you may make predictions. Because of many prejudices it is easily ignored. In that sense Darwin's principle of natural selection is equally trivial. In spite of this it had to be pushed through much psychological opposition."

The matter is different in the case of the strong version of the principle. Critics claim it to be tautological: the world is as it is because it is as it is. In fact, this hypothesis reaches much deeper into the structure of the physical world and hence may be questioned more strongly than the weak version. Primarily it claims the physical properties of the universe — its fundamental parameters and coupling constants — must be restricted in order to be compatible with life as we know it. (This logic could be employed by anyone in the universe; this is only "anthropic" as long as only we do it; but it is not anthropomorphic at all!) "Of course such a principle is only of value," Brandon Carter comments, "if these restrictions are sufficiently strong. It hardly is good for predictions. But if it is true as is indicated by certain fine-tunings in the laws of nature, then at least this principle will serve to clarify and coordinate. After all, science has not only its value in the predictions of things one does not know about. It also serves a purpose — and here biology is a special example — to bring order into the multitude of facts which one already knows about."

This way, there is no danger of our over-valuing the anthropic principle. It is a hypothesis dressed in the methodology of a principle, which places Man, as the cognitive medium of a universe, with surprising regularity at the focal point, the hub around which the laws of physics operate.

None of this is changed if, as is now popular, the tables are turned and we are told that this and that could have been predicted. The negative side of this discussion is our lack of knowledge as regards extraterrestrial life: if we knew more, we could understand earthly life better too.

18

According to our present level of knowledge, life as we know it can only function on the basis of chemical elements heavier than hydrogen and helium. This demands the existence of galaxies and special types of stars and planets. It is however conceivable, though not particularly probable, that a different form of intelligence could get by without any of these ingredients.

In explaining "accidental" properties of reality, the anthropic principle cannot claim the precision of other, physical explanations; it does not give the precise mathematical values of the fundamental constants and mass relationships, but generally establishes their orders of magnitude. "With enough anthropic conditions," the British physicists Bernard Carr and Martin Rees note, "one may be able to be more precise about the constants of nature, but the present situation is unsatisfactory."

Even with the assistance of the anthropic principle, some cross-connections between constants and mass relationships still remain unaccountable or coincidental. It is possible that one day we may discover "ordinary" physical explanations for those observable coincidences the anthropic principle is unable to explain; but "even if all apparently anthropic coincidences could be explained in this way," Carr and Rees remark, "it would still be remarkable that the relationships dictated by physical theory happened also to be those propitious for life." For the moment it is an open question whether the anthropic principle is no more than a philosophical curiosity or whether it embodies some profound truth about life and the universe of which we have as yet lifted only a corner of the veil.

Where might a consistent and successful application of the anthropic principle take us? Lines of research based on the anthropic principle may eventually form part of a wider program — not a theory directed to the more conceptual-philosophical "unity of nature," like that of Carl Friedrich von Weizsäcker, but, as proposed by Dennis W. Sciama of Oxford and Edward R. Harrison of Amherst, centered on the "unity of the universe." Every part of the universe is in a — more or less direct — relationship to all its other parts, and all influence the rest; and this justifies the hope that we may one day be able to understand the entire universe by examining only one part, the part in which we live, which is in any case the limit imposed on us as observers tied to the Earth. The Earth and the observers of the cosmos living on it would then be so firmly anchored in the overall structure of the cosmos that we could comprehend all of existence through the exploration of our tiny corner of space alone. In principle at least we will one day achieve this knowledge, if — to adopt George F. Ellis' formulation — "there is ultimately only one self-consistent structure of a universe in which we can live."

Chapter II
Dirac and
Cosmic Coincidence

It must have taken Paul Dirac's breath away when, around 1937, he turned from his concern with the structure of the microcosm to the macrocosm in the form of the universe at large. Although established as Professor of Mathematics and Physics at Oxford and joint holder, with Erwin Schrödinger of Vienna, of the 1933 Nobel Prize for Physics, he was, as far as cosmology was concerned, an outsider, a newcomer. His contemporaries such as Eddington, Jeans, Hubble, Milne and others were already ensconced as the accepted authorities in the field. But the very status of an outsider will occasionally, though not always, open the way to a fruitful new perspective, a fresh insight.

The magic of cosmology has always been in the huge numbers demanded by a description of the skies — whether this involves a simple count of the stars or Archimedes' calculation of the number of grains of sand which would fit into the universe. Perhaps it was a fascination with just this aspect and a playful curiosity that led Dirac to turn to cosmology. In any event he very soon stumbled upon a remarkable discovery. Taking up earlier number games which had occupied Eddington, he brought a number of facts together.

Fact Number 1: In giving the age of the universe, the terrestrial year may be a comfortable unit, but in physical terms it is not a particularly rational measure of a period running to fifteen or twenty billion years. Nature, on the other hand, has created in the electron

21

and proton of the atom a precise clock and a scale of time which will measure the universe in a far more sensible and natural way than the period of the Earth's revolution around the sun. Counted in these atomic time units, the universe today is 10^{40}, that is, 10,000,000,000,000,000,000,000,000,000,000,000,000,000 units old.

Fact Number 2 : If we calculate the reciprocal of the gravitational (fine structure) constant, we once again arrive at 10^{40} (see table on page 33). If we then turn to the number of particles contained in the observable universe, the result is — within an order of magnitude — a closely related figure: the square of 10^{40} — that is, 10^{80}, or $(10^{40})^2$.

The figure 10^{40} has now cropped up three times. Is this all just a coincidence, the product of some irrational magic of numbers? These large numbers may be strikingly similar, but they stem from totally separate branches of science: the age of the universe is a cosmological matter; the gravitational constant measures the strength of gravity; and the atomic time unit is an element of the microcosm. For want of any ready explanation for this uncanny correspondence, it was simply referred to in terms of "cosmic coincidences." But Dirac was unwilling to accept them as coincidences, for in his opinion there were no coincidences in nature. Rather, these "coincidences" must, he felt, imply some deeper connection between the largest and the smallest phenomena which still awaited elucidation. Dirac therefore tried to reformulate his cosmic coincidences as scientific necessities. We will come back to this later.

PHYSICALISM AND THE NATURAL WORLD

Dirac quickly overcame his surprise and — as befits a great theoretician — drew up the foundations of a new theory. But what may well have been easier in 1938, that is, pulling a new theory out of a hat, as Einstein used to do, is getting harder as time goes on. We already have reasonably well-functioning theories for many kinds of natural phenomena, while for others we have no explanation, or we are cautiously feeling our way forward through problems that are still too great for us. But Dirac's "discovery" of the cosmic coincidences and the attempt to explain them by means of the anthropic principle have now set us a clear task. These "anthropic" cross-connections (since that is what we shall prove they are) do not compel us to set up a new theory; but there is no doubt that they point to gaps in the coverage of existing theories. We must therefore ask ourselves: Is our stock of natural laws sufficient to make the existence of *homo sapiens* a "physical necessity" — the inevitable result of a cosmos shaped like the one we know?

Here we have two choices. One would be to dismiss Dirac's "discovery" as an irrelevant numbers game. This would correspond to an attitude widely found among physicists who are happy to accept many independent natural constants and several fundamental interactions equal in status without feeling the need to insist on any hidden connection between them. The other would be to explain it by some all-embracing, total theory of — perhaps, at bottom, inanimate — natural processes. The drawback of the first response is that we would be turning down any chance of finding a possible scientific explanation. For this reason, the second alternative will form the basis of this study. It leads us directly to two fundamental questions of natural philosophy:

— Are our theories of nature in any way complete, or should we, for that matter, expect that they ever will be?
— Are there areas of nature, biology and life which in principle *cannot* be put on a foundation of physical laws?

Behind these questions lurks a historical argument, the so-called "physicalist controversy." The thesis of physicalism is that all processes and properties can indeed be seen as consequences of physical laws. As, even in biology, these are in the end reducible to the actions of matter and radiation, this view is entirely plausible, and therefore suits the convictions of most scientists. The controversy has broken out again in recent years over the question of artificial intelligence and its comparison with the human brain. Is the human brain capable of a full understanding of itself? A related question is whether we will ever be able to build a computer that will be just as "intelligent" as a human being. For many "intelligent" human actions, the answer would be "yes." But opinions are divided when it comes to the qualities considered specifically human, such as creativity and the capacity to have feelings, desires, and longings.

23

How human-like computers will ever become is something we do not know; but a little skepticism is not out of place. A chess computer no human player can beat is still no more than a complex but relatively stupid bundle of microelectronics.

The second point concerns the completeness of physics. First, a basic consideration: *every* theory within science is valid only for a limited segment of reality. These theoretical limits, however, are not part of the theory and are not contained in it. The theory is only shown its limits by a new, wider theory that also contains the old theory within it (as a special application) or borders on it. As a result, terms and concepts connected with the older theory are "superseded" by analogous elements in the newer one. This does not mean that the old concepts are no longer of any use — quite the contrary. Werner Heisenberg, who originated the concept of the "closed theory," says: "Even when the limits of the 'closed theory' have been broken, that is, when new fields of experience have been organized with new terminologies, the conceptual system of the closed theory remains an indispensable part of the language we use to talk about nature. The closed theory is thus one of the necessary foundations of further research; we can only express the result in the terminology of earlier closed theories." According to Heisenberg, the characteristics identifying a closed theory are its lack of internal contradictions and the fact that it represents experience as well as it is possible to do so. He summarizes its truth content in the following theses:

> "The closed theory is valid for all times; wherever experience can be recorded in its terminology, even if it be in the remotest future, its laws will always prove correct."

> "The closed theory contains no completely certain statement concerning the world of experience. For it remains in the strictest sense uncertain and simply a matter of success and failure to just what extent phenomena can be comprehended with the concepts of the theory."

24

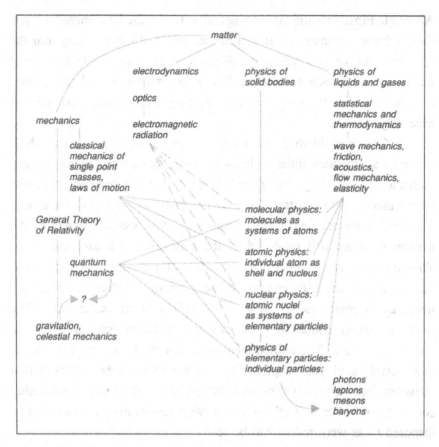

From the macrocosm to the microcosm, the various disciplines of physics describe the behavior of matter and radiation. [Freely redrawn, after E. Lüscher, H. Jodl (eds.): Physick - Einmal Anders (München, 1971), vol. I, p. 73).

> "Despite this uncertainty, the closed theory remains
> a part of our scientific language and thus forms an
> integrating element of our understanding of the world
> at any given time."

This makes it reasonable to suspect that there will never be any final closed theory. No theory can hold unlimited validity if its field of application remains unknown to us. This could even be the case with a unified theory intended to take in all four fundamental interactions.

Although Einstein's dream may one day be fulfilled by means of just such a theory, chances are it will neither be an absolute theory, nor the last one. As we continue advancing into ever more exotic reaches of reality — still greater energies, still smaller dimensions and still shorter time intervals — these new fields may continually make new theories necessary.

A further problem arises from the fundamental constants, which occur in all theories hitherto known. Their occurrence in any theory marks a measure of ignorance regarding the area of reality encompassed by the theory. One constant, for instance, gives the strength of a natural force, but does not reveal the reasons why it is just that strong or the manner in which it is produced by the interaction of matter and fields. Thus every constant conceals a fragment of the natural world that is not yet explained, and remains in itself simply a "phenomenological" item: an external description with no self-contained explanation. The fewer constants contained in a theory, the better we can claim to understand nature. The unified theory that scientists are currently hoping to see established (see Chapter III) will reduce the number of interaction constants from four to two or perhaps only one. Whether this remaining constant can be "explained" by the unifying super-force, and can thus be disposed of as well, remains to be seen.

This is where the task of science comes up against the boundaries of philosophy and religion. It can only set itself the aim of deriving structures and consequences from general and wider contexts. Explaining nature as a whole is beyond its ability and in any case outside its responsibility. It will not, therefore (runs the prevailing view), be expected to explain why the laws of nature are just the way they are. So conventional science stays within the modest gambit assigned to it — and is all the more successful for this! Taking things pragmatically, it does not set out to explain the science that is being practiced. On this basis, a theory containing phenomenological constants will be no "worse" — except in an aesthetic sense — than one devoid of such constants, as long as it correctly describes the processes that lie in its area of application.

At this point the anthropic principle joins the action. With its help two lines of thought in particular can be followed up:

— What is special about the time and place of human existence, and what cosmic circumstances and evolutionary processes must have prevailed; and

— what structure had to be established by nature in order that the human race could come into being at all.

To this extent, the strong anthropic principle aims to "explain" why the laws of nature are the way they are, even though the variety of "explanations" offered by the anthropic principle — a teleological one — is of course of quite a different kind from the causally-organized explanation given by physical theory. What we can do with the weak anthropic principle, however, is use the singular phenomenon of human existence to deduce certain chains of causality. This, of course, will be done within the framework of an existing theory; the physical "explanations" are still provided in this case by the theory itself, not by the anthropic principle. But these deductions are the quintessence of the anthropic viewpoint — if they make it plausible that a cosmos with different natural laws would not permit an intelligent life form to develop.

Natural forces and Their Fundamental Constants

Scientists concerned with the fundamental structure of the natural world tell us that all processes in the cosmos are controlled by four fundamental forces (or, as they prefer to call them, interactions). Only four forces? This sounds very straightforward — too simple, in fact, if we consider the enormous variety of complex formations and structures we meet within nature: stars, galaxies, planets, chemistry and the world of biology, through to ourselves and our brains.

Today's "belief" in the four fundamental forces may remind some people of the long-dead picture of the world held by the pre-Socratic philosophers of classical Greece. At that time, the world was thought to be built up from four "elements": fire, air, Earth, and water. Two-and-a-half millenia later, our "elements" amount to over one hundred. Could the version of nature offered by today's physics be just as "naive" in its own way as that of over two thousand years ago? Can we be sure that, a hundred years or more from now people will not look back indulgently on the science of the late twentieth century, commenting with the wise and impatient sigh of those who come after: "Well, of course, in those days they didn't know about...?"

The reason successive generations can see further than their scientific forebears is that, in the time-honored phrase, they are standing on the shoulders of giants — though it may only be dwarves that we find perched atop their predecessors. We certainly would not yet claim that our picture of the physical world is complete in every detail; but on the other hand, it has certainly lost much of the naiveté that characterized, for instance, the mechanistic cosmology of the nineteenth century. It is hardly to be expected that in fifty years' time we will suddenly be talking of not four, but some dozens of different forces. In describing the world familiar to us today, at least there will be no need to give up the four fundamental constants in a hurry. Perhaps we will have five, maybe one unified force. But at this stage it is unlikely that a description of nature will require more than four (or five) forces (see Chapter III). This is supported by the fact that we understand rather better than we did ten years ago how nature can contrive such a complicated universe with such "simple" ingredients: matter and radiation, coupled only by four types of force. The most obvious characteristic of this structure, and at the same time of primary significance in our study, is the fact that the fundamental forces differ in strength. It is on the basis of the different strengths of the four forces that scientists are able to offer a plausible explanation of how the variety in nature came about. And the converse also applies: it can be demonstrated that, if the forces are all alike in strength, this variety disappears and gives way to uniformity (see Chapter IX).

I propose to show in the following pages how little (or how much) we can "adjust" these forces without losing the variety which includes our own existence — or, more precisely, how little room for change is allowed by the criterion of humanity's existence.

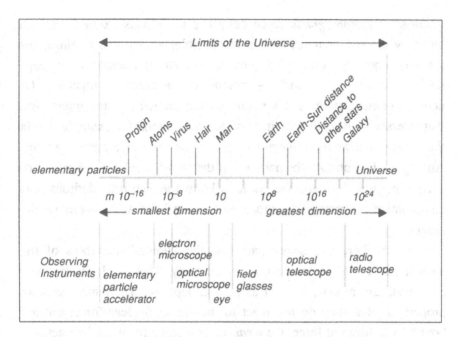

From its smallest to its largest dimensions, the natural world can be observed with the appropriate instrument. All objects are shaped by the operation of only four forces. In the center is Man.

Which forces, then, are we dealing with? In order of increasing strength, they are as follows. First, the force of gravity, which dominates interactions between larger, macroscopic bodies. After this come three forces controlling events in the microcosm: the weak nuclear force, the electromagnetic force, and the strong nuclear force. Instead of "forces," physicists also speak of "interactions," a designation chosen for the simple reason that the influence of any body on another is matched by an influence of the second on the first. As a law of mechanics, this was formulated by Isaac Newton as the principle that "every action has an equal and opposite reaction." Today, the term is applied more generally to all effects that bodies and fields can exert on each other.

Gravitation regulates the influences of astronomical bodies on each other and decides the motions of planets, stars, galaxies, and the entire

29

cosmos. *Electromagnetic force* describes the effects exerted on each other by electric charges, and is responsible, among other things, for the way atoms and molecules interact. The *strong nuclear force* keeps nuclear particles (nucleons) — protons and neutrons — together. (A note on nomenclature at this point: all the particles of the microcosm that interact — in collisions in particle accelerators, for example — via the strong nuclear force have been given the collective name *hadrons*. Among the hadrons are the particles of the atomic nucleus — *baryons* or "heavy particles" — and the particles known as *mesons*. Hadrons thus compromise neutrons and protons on the one hand and mesons on the other).

Where there are heavy particles, the "lighter" members of the particle family, the *leptons*, are never far away. These include electrons, neutrinos, and muons, as well as heavier leptons. Their most striking property is that they do not react to the strong nuclear force, but are bound by a different force, the *weak nuclear force* (or weak interaction). This can be observed, for example, in radioactivity, during beta decay (beta particles are electrons). During beta decay a neutron in the nucleus of an atom will split into three parts: a proton, an electron, and a neutrino (or an anti-neutrino). In this process the proton remains behind in the nucleus; it is bound to the other nuclear particles by the strong nuclear force. But the atomic nucleus now contains one proton charge more than before, and this creates a new chemical element. Meanwhile, the electron and the neutrino leave the nucleus and, as leptons, are disregarded by the strong nuclear force.

The difference between all these forces — that is, the relative strengths of the four — can be most simply expressed by numbers, the "physical constants" typical to the respective interactions. The weak force outweighs gravity by a factor of 10^{28}; electromagnetic force is 10^9 times stronger than the weak force; and the strong interaction outweighs electromagnetic interaction by one hundred times. Adding all these up, gravitation and the strong nuclear force are separated by 40 orders of magnitude, a relative factor of a full 10^{40}! (The same number that Dirac was investigating).

Given this wide diversity in the strengths of the springs which make the clocks of the universe tick, there is no room for confusion as to which of the four forces is principally at work in any particular process. Moreover, they differ not only in their strength, but also — and equally dramatically — in their respective ranges. The rule of thumb here is: the stronger the force, the more limited is its range. The nuclear forces, such as the strong interaction, first described as long ago as 1935 in the work of the Japanese physicist Hideki Yukawa, only operate over the diameter of a proton — that is, across just 10^{-13} centimeters.

FORCE	STRENGTH	RANGE	PARTICLES AFFECTED	PARTICLES EXCHANGED	ACTION OF FORCE BETWEEN SIMILAR PARTICLES
Strong	1	short	quarks	gluons	attractive
Electro-magnetic	10^{-2}	long	electrically charged particles	photons	repulsive
Weak	10^{-11}	short	electrons, neutrinos and quarks	intermediate vector bosons	repulsive
Gravity	10^{-39}	long	all particles	gravitons	attractive

The four fundamental forces of nature and their most important properties

(How short a distance this is may need illustration. A proton is itself only one hundred-thousandth of the size of a hydrogen atom; and 10^{-13} centimeters stands in the same relationship to 1 centimeter as 1 centimeter does to the distance from Earth to sun.)

Electricity and gravity behave in a quite different way. In principle, both have infinite range; but the electrical force is usually neutralized very quickly and over a short distance. Large bodies contain about as many positive as negative charges, so that electrons (with negative charge) and protons (with positive charge) always cancel out each other's charges in larger macroscopic bodies; thus large bodies are in the main electrically neutral, with the result that the electromagnetic force mainly operates in

31

the microcosm. And in the realm of elementary particles, gravity in its turn has no effect — it is far too weak. It is restricted to greater distances and therefore rules the macrocosm. Gravitational force increases with the amount of matter, for there is only one kind of "gravitational charge," the mass of bodies, and nothing can cancel out gravitational attraction. So it comes about that the macrocosm of celestial bodies is controlled entirely by the weakest of all forces.

It will help in discussing the anthropic principle and the fundamental constants if we give the different constants handy names and refer to them as *alpha strong, alpha electromagnetic, alpha weak,* and *alpha gravity* — or, in abbreviated form: alpha−S, alpha−E, alpha−W, alpha−G (alpha-frequently occurs in formulae as the symbol for one or other of these forces). The precise numerical values are: alpha−S = 3.9; alpha−E = 0.007; alpha−W = 10^{-11}; alpha−G = 0.5×10^{-40} (see Table V in the Appendix).

Given in this form as dimensionless numbers, these constants have become known to scientists as "fine structure constants" — thus, alpha−E is the electromagnetic fine structure constant, alpha−G is the gravitational fine structure constant, and so on.

These figures, from which the relationship between constants, mentioned above, can easily be worked out, must in principle – taken together with the relevant natural laws — hold the secret of life. Following this lead and making the connections is what will occupy us in the next few chapters.

We may justifiably ask how it is that these numbers came about, and why they have these values and no other. They certainly did not fall from the sky of pure thought. The excursion into "dimensionless numbers in physics" stimulated by their mention is a good way of learning how theoretical physicists think. In a word, *these dimensionless constants are the only ones we are concerned with* (rather than other quantities with a dimension, such as one gram or five centimeters) if we are considering the possibility of physically changing such "constants." As Robert H. Dicke stressed in 1962, it doesn't mean a lot if we simply speak of changing the mass of an electron as expressed in the dimension of grams; instead, we must turn to the relationship between two masses, such as the ratio of the mass of the electron to that of the proton, or that of the electron's mass to that of the proton, or that of the electron's mass to the standard Paris kilogram. Only ratios like these, dimensionless numbers, are free of the arbitrary units fixed by man. As we are free to make a change in the Paris kilogram from one day to the next, we could by so doing conjure up apparent changes in the natural constants which would, of course, make no change in the reality of the physical world, and which would therefore impress no one. But this cannot happen to the ratio of electron mass to proton mass (=1/1837); this figure is independent of Earthly conventions deciding how big a gram should be.

It should not surprise us, then, that the four fundamental constants of the natural forces are only "dimensionless" through their being built up from other, dimension-bound constants — that is through the appropriate multiplication and division of other fundamental and more routine quantities of physics. These are:

— the speed of light;
— Planck's constant, or "quantum of action" (one of the fundamentals of quantum mechanics theory);
— the mass of the proton;
— the mass of the electron (about two thousand times smaller than the proton mass);
— the charge of the electron;
— the gravitational constant (as introduced by Sir Isaac Newton in 1666 in his Law of Gravity);
— the weak interaction constant; and
— the strong interaction constant.

The strength of the four forces of nature is contained in the last four of these quantities, which, however, must first be converted into dimensionless numbers. This is easily done with the help of the first five atomic constants in the list (see also table on p. 31).

33

THE FUNDAMENTAL CONSTANTS OF MATTER
AND THE MASSES OF THE ELEMENTARY PARTICLES

After the natural forces we must turn to the material constituents of the universe and understand their role in the clockwork of nature and as the structural basis of life. Particularly when dealing with the late stage of cosmic development in which we live, and especially the fate of the Earth, on which we dwell, we find matter — that is, above all else its stable components — playing a more decisive role. It is the elementary particles, the smallest known building blocks of matter, that are the components of atomic nuclei and the atoms themselves — which in turn are, of course, what we are made of.

For every particle there is an anti-particle of equal status. These make up the infamous *anti-matter* — "infamous" because the encounter of any particle with its anti-particle is fatal to both. In their mutual destruction they radiate electromagnetic rays of extremely high energy, known as gamma rays. It is as a result of this incompatibility that stable, inactive matter, such as we are made of and have to deal with, contains no anti-matter. (Conversely, a stable construction of anti-matter contains no matter).

It is also a consequence of this incompatibility that no way has yet been found to put this radiation process to military use. It is by no means a simple task: anti-matter can only be produced at great effort, and it is even more difficult to store it in our own environment. This is why the deployment of "anti-matter guns" has so far been restricted to science fiction.

As late as 1932 it was still believed that matter was constructed from just two elementary particles — the electron and the proton. But although there is now a veritable menagerie of particles, with several hundred "exhibits" to keep physicists on their toes, only a small selection of these is of significance here, because the vast majority of elementary particles discovered since 1932 are unstable and can only be produced using very large amounts of energy in accelerators measuring kilometers across. They are correspondingly short-lived, too, with lifetimes ranging from a few minutes to tiny fractions of seconds.

34

As far as stable matter is concerned, the circle of relevant building blocks has not significantly increased since 1932. Here, only the following elementary particles have a part to play: the particle of light energy (photon), the neutrino (the lepton that appears in beta decay), and our old friends the electron, proton, and neutron. For terrestrial matter, the matching anti-particles can be disregarded, since conditions in the Big Bang led to a state in which the part of the universe we know can be seen to consist almost exclusively of matter, with no appreciable measure of anti-matter. A trivial exception, however, is the photon: this is the one elementary particle which is at the same time its own anti-particle. Or, to put it another way, as a massless particle and the long-range medium of electromagnetic force, it has no anti-particle. The neutron, left to itself, will decay in an average time of 15 minutes; only when it is "packed" into an atomic nucleus does it survive much longer. The remaining four — stable — elementary particles cannot decay by themselves or convert spontaneously into other particles; their life expectancy can thus be said to be "infinite." In the case of the proton, however, the question of stability remains an open one: it is possible that it does decay after a very long time (see Chapter III). But this possibility, which is no more than a theoretical one as yet, does not bear on our terrestrial existence, since the timescale in question is many times the age of the universe.

THE FUNDAMENTAL PARAMETERS OF THE COSMOS

The third group of fundamental quantities that are decisive for our universe in both microcosm and macrocosm — after the constants and the elementary particles — apply to the cosmos itself. Here it is a question of establishing relationships between properties of the cosmos on the one hand and properties of microphysics on the other, to see how far these relationships tie in with the anthropic principle. The question we start with, therefore, is: which physical quantities decide the nature of the cosmos, at least in its overall behavior? Without going into a description of the models cosmologists are currently working on (see Chapter IV), it is enough to say that the overall structure of the cosmos is determined by remarkably few factors — more precisely, by a few

35

parameters. Parameters are variables, whose values are differently fixed in different models. These are:

— The Hubble constant. This puts a figure on the speed of recession (movement away from us and each other) of distant galaxies. For every light year away from us, the relative recessional velocity increases by about 1 centimeter per second. The reciprocal of the Hubble constant provides a rough figure for the age of the universe: the current value is about fifteen to twenty billion years.

— The mean density of matter in the universe. This is presently estimated as being on the order of a few dozen nuclear particles per cubic meter, corresponding to 10^{-31} grams per cubic centimeter. This estimate is today taken as a lower limit and could in fact be considerably increased, as astronomers are unable to be sure how much matter is bound up in invisible objects (such as black holes and neutrinos).

From the Hubble constant and the mean density of matter we can calculate two further cosmic numbers, if we first postulate a particular model of the cosmos to use as a basis:

— The "radius" of the universe — a rather imprecise, but readily visualized concept giving a linear scale typical for the expanding universe. (The "radius" of an open cosmos would be infinite.) In a similarly imprecise but graphic way, we can speak of the "volume" of the universe — that is, the capacity of sphere with the "radius" of the universe — and a total mass, reached by multipyling the volume by the mean density of matter.

— The number of heavy elementary particles (hadrons), such as protons, in the universe.

The anthropic principle can now provide us with a method of examining the construction and underlying foundations of the universe, and indeed of the entire material and immaterial cosmos, so as to find out what is special about it. What may come out of this is the distinctive content of the real universe, and perhaps a certain singularity and uniqueness, the more the existence of an intelligent life form narrows down the possible variety of its construction. The end result may then be statements on the uniqueness of the human race, the uniqueness of life, the uniqueness of the physical world and the uniqueness of the cosmos.

The Classical "Cosmic Coincidences"
— An Expression of the Anthropic Principle?

Since the end of the fifties, Paul Dirac's "Large Number Hypothesis" has opened a Pandora's Box of great fascination to physicists, and the discussion is still far from over. With his theory, Dirac brought something to life which is still scorned as numerology by many theoreticians. Dirac compelled scientists to break with their accustomed habits and turn away from their enforced, but often mindless, obsession with details to consider connections which cover the whole spectrum of science without regard to the boundaries between disciplines, embracing fields from atomic physics to macrophysics and eras from the Big Bang to the evolution of intelligence. For the physicist, this is a new way of exploring the "unity of the physical world"; and it satisfies the researcher's old yearning to scan and catalogue the universe by means of the simplest possible scheme. This is a delicate matter for a scientist, as he is forced to cross the immediate frontiers of his specialty (naturally harvesting the inevitable criticism in the process); but it can be important too, because the philosophy underlying all scientific endeavor makes for a conviction that in the end will be possible to bring all the activities of the cosmos — from the Big Bang to mankind's existence — under one theoretical roof.

The terms "cosmic coincidences" or "large number hypothesis" refer to a set of numerical cross-connections between the fundamental constants of microphysics on the one hand and the fundamental parameters of the cosmos on the other. These can be illustrated and "interpreted" in several different ways.

The large number that caught Dirac's eye was 10^{40}. This seems to crop up in quite a variety of contexts. Let us take one fact as a beginning: the reciprocal of the gravitational constant (alpha-G, the gravitational fine structure constant) is about 10^{40}. Then we have

		Power of N ($N = 10^{80}$)
	$1/(\text{alpha–G})$	$N^{1/2} \sim 10^{40}$
Coincidence No. 1:	$\dfrac{\text{"Age" of the universe}}{\text{electron timescale}}$	$N^{1/2} \sim 10^{40}$
Coincidence No. 2:	Number of particles in visible universe	$N \quad \sim 10^{80}$
Coincidence No. 3:	$\dfrac{\text{electrical force}}{\text{gravitational force}}$	$N^{1/2} \sim 10^{40}$
Coincidence No. 4:	Gravitational radius of universe	$N^0 \sim 1$
Coincidence No. 5:	$\dfrac{\text{"Size" of electron}}{\text{Planck length}}$	$N^{1/4} \sim 10^{20}$
Coincidence No. 6:	$\dfrac{\text{Number of photons in universe}}{\text{Number of baryons in universe}}$	$N^{1/8} \sim 10^{10}$

The large numbers of nature and the cosmic coincidences

Cosmic Coincidence No. 1: The "radius of the universe," a linear scale characteristic of our cosmos, is around 10^{40} times the classical electron radius.

To put it another way: the ratio of the radius of the universe to the radius of the electron is 10^{40}.

If these distances are replaced by the time taken by light to cover them, we can just as easily express Coincidence No. 1 in periods of time. Instead of the radius of the universe we would then take its age, and instead of the radius of the electron the "atomic time unit," known to physicists as the electron timescale — that vanishingly small time of 10^{-23} seconds that light requires to travel a distance equal to the classical electron radius. Here, the age of the universe is equal to about 10^{40} times the electron timescale. If we imagine a watch that ticks once every 10^{-23} terrestrial seconds (one hundred thousand billion billion times every second), it would have ticked 10^{40} times between the Big Bang and now.

Cosmic Coincidence No. 2: The universe contains within the boundaries we can see about $10^{40} \times 10^{40}$ or $(10^{40})^2$ heavy elementary particles such as the proton — or, in short, the mass of the visible universe is about $(10^{40})^2$ proton masses.

An alternative formulation of one of the cosmic coincidences can be illustrated by means of the hydrogen atom. In this atom, one electron orbits one proton. The two particles having matching but opposed electric charges, and different masses. The electron is some two thousand times lighter than the proton. Proton and electron attract each other electromagnetically through their charges, but also through gravity acting on their masses. If we compare the electrical and gravitational forces acting between the two particles, we find

Cosmic Coincidence No.3: The electrical attraction between electron and proton is 10^{40} times as strong as the force of gravity between them.

And yet a fourth cosmic coincidence: every particle in a gravitational field possesses potential energy. If the particle is allowed to drop, like an apple falling from a tree, this potential energy is converted into kinetic energy (energy of motion). In the cosmic gravitational field the particle also possesses a certain mean potential energy with respect to all the other matter in the universe. (For simplicity, we may imagine the universe as a sphere with the radius and mass that have been calculated for it, as mentioned above.) This potential energy may be compared with the rest mass energy of the particle under consideration, as given by Einstein's formula $E = mc^2$. This brings us to

Cosmic Coincidence No. 4: The potential energy of a particle of the cosmos is about the same as its rest mass energy.

The ratio of one to the other — (cosmic potential energy) / (rest mass energy) — is thus 1. This figure — which is also, of course, a power of 10^{40}; it is $(10^{40})^0$ — corresponds to the "gravitational radius" of the universe. This is the radius of a hypothetical single body in the form of a black hole with the same mass. Two further coincidences are listed in the table.

To sum up: in all these cosmic coincidences we find the colossal figure of 10^{40} ominously recurring — a circumstance which has continually provoked scientists' curiosity. Paul Dirac's conclusions were unequivocal. In 1938 he drew up the following postulate: "any two of the very large dimensionless numbers occurring in nature are connected by a simple mathematical relation in which the coefficients are of order-of-magnitude unity."

Dirac rejected the suggestion that he might simply have recorded the workings of chance. And it may have been the cosmic connection between the age of the universe and the atomic unit — Cosmic Coincidence No.1 above (second version) — that suggested the following idea to him. In a developing universe the age of the universe cannot be a "constant, as it changes with the march of cosmic time." Perhaps all the large cosmological numbers likewise undergo change; in particular, the gravitational constant alpha$-$G $= 10^{40}$ might change as cosmic time goes by. Dirac constructed his own cosmological theory on this hypothesis (see Dirac's cosmology in Chapter IV).

However, a link between cosmic coincidences and life on Earth was only postulated twenty-four years after Dirac by an American scientist. In 1961 Robert H. Dicke of Princeton proposed a remarkable solution to the riddle of the large cosmic numbers. At the very least, he said, the cosmic coincidences should have come as no surprise. In the cosmic past, when the ratio between the age of the universe and the electron timescale was still less than 10^{40}, there was no one around to comment on the fact.

Intelligent life could not arise before the first stars had created in their interiors enough of the heavier elements necessary for the development of biological systems. This period of time, however, corresponds to the present age of the universe. On the other hand, in a later epoch, when the cosmic large numbers had once again altered, there would only remain a few stars, and these would be low in energy, so that once again life would be unlikely to come about and perpetuate itself in their vicinity. In other words, the discovery of 10^{40} as a special cosmological number tells us no more than that the human race, an observer of the universe, can exist. Or, even more succinctly: we exist because the universe is as old as it is.

This argument by Dicke may be regarded as the beginning of serious discussion of the weak anthropic principle. The age of the universe is a prediction based on this principle.

Dirac, however, had another answer ready. In the same issue of the *Physical Review* as Dicke's article, he wrote:

"Dicke discusses the three cosmological numbers: (1) which determines the gravitational constant, (2) which determines the Hubble age of the universe, and (3) the number of particles in the universe. They are related in that (1) is the reciprocal of (2) and (3) is roughly the square of (2). I asssumed that these relations correspond to something fundamental in nature. With an evolutionary universe (2) varies with time, and the (1) and (3) would also have to vary with time.

"Dicke ... shows that (2) would have to have roughly its present value when habitable planets exist. On this assumption habitable planets could exist only for a limited period of time. With my assumption they could exist indefinitely in the future and life need never end.

"There is no decisive argument for deciding between these assumptions. I prefer the one that allows the possibility of endless life. One may hope that some day the question will be decided by direct observation."

Dirac's objection is not only subjective but false as well. Dicke's association of the existence of life and the age of the universe does *not*, of course, exclude the possibility that life, not necessarily biological, might also exist much later, even at a time (on the cosmic timescale) when the

41

large cosmic numbers might have shifted to quite different values. True, the birth of a new species will have become very much less likely by then. Most stars will have died, and those remaining will be radiating very little energy and even so at unfavorable wavelengths. On the otherwise so fruitful Earth, a star too weak in energy at the life-giving wavelengths would never get plant photosynthesis started to any appreciable extent. The weak anthropic principle gives us an approximate value for the age of the universe: before the present cosmic epoch there simply could not be any observers to stumble upon the cosmic coincidences; and later, with an ever-decreasing number of sun-like stars available, the chances of life coming into being and surviving the tedious biological processes of evolution become slimmer and slimmer.

Who Was the First?

Paul Dirac, in 1938, was the first to attempt an "explanation" for the cosmic coincidences, but he was far from being the first to have wondered at the large numbers of nature. The earliest speculations of modern times concerning the possible significance of large cosmic numbers date from 1919 and are those of the mathematician, physicist and philosopher Hermann Weyl, who was then teaching at the National Technical University of Zurich (Switzerland) after leaving the German university of Göttingen.

Weyl was followed in his preoccupation with the large numbers by the British theoretician Sir Arthur Eddington. The square of 10^{40}, 10^{80}, is often referred to as "Eddington's number" after him. Then came Dirac, followed in turn by Robert H. Dicke, the Hamburg (Germany) theoretician and co-founder of quantum mechanics Pascual Jordan (died 1980), Oskar Klein, George Gamow, the "inventor" of the Big Bang, and a number of present-day scientists who also fall prey to the lure of the constants and the large numbers game.

There is also, however, one record of scientific concern with large numbers in antiquity. Remarkably enough, it seems that an attempt was made to calculate the number of particles in the universe two thousand years before Eddington — by the Greek mathematician Archimedes who lived in Syracuse from 287 to 212 BC. His work, "The Sand-Reckoner," which he dedicated to his ruler, King Gelon, begins with the words: "There are some, King Gelon, who think that the number of sand particles is infinite in multitude; and I mean by the sand not only that which exists about Syracuse and the rest of Sicily but also that which is found in every region whether inhabited or uninhabited. Again there are some who, without regarding it as infinite, yet think that no number has been named which is great enough to exceed its multitude."

In the third century BC, the Greeks employed a system of counting in which they could count up to the square of a myriad – 1 myriad being 10,000. A myriad squared thus equals $(10^4)^2 = 10^8$ or 100 million. Up to this figure, then, the Greeks could count without complication. Figures in excess of 1 myriad squared could still be stated, but in a rather more involved way. Archimedes, who clearly considered a grain of sand to be the smallest particle of mass in nature (or at least thought it a widely distributed unit of mass which could usefully stand in), wished to calculate the maximum number of grains of sand that could be accommodated by the sphere of fixed stars, in accordance with the model of the universe prevailing at the time. (The cosmology of ancient Greece considered the stars to be attached to crystal spheres that rotated in opposing directions, all with the Earth at their center.) After complex calculations Archimedes arrived at a figure for the number of grains of sand in the universe: written in the notation we use today, his estimate is 10^{63}. It is fair to point out, however, that Archimedes was working with extremely small grains of sand; he assumed a poppy seed to be "one-tenth of a finger" wide, and that this could contain a myriad grain of sand.

With a little skill, Archimedes' grains of sand can be replaced by "modern" fundamental particles, e.g. protons. It is a simple conversion. The width of a finger equals about one centimeter, so one-tenth of this is one millimeter; sand grains are about three times as heavy as water, giving a density of 3g per cc. With 10,000 grains of sand in a poppy seed, there are 3×10^{-7}g, or around 10^{17} nuclear particles to each grain. If this last figure is multiplied by Archimedes' figure for the total number of sand grains, 10^{63}, this gives a final figure of $10^{63} \times 10^{17}$, or 10^{80}, as the number of particles in the (Archimedean) universe. By a happy accident (?), this exactly matches the number of particles assumed to exist in the matter of the universe today!

Archimedes, however, was doing no more than making an educated guess — for, by today's standards, he underestimated the size of the universe by a considerable amount. His assumed distance to the crystal sphere holding the fixed stars, about one-and-a-half billion kilometers, would just include the orbit of Jupiter. If Archimedes' calculations were repeated with today's data, the resulting figure would be far too large: setting the radius of the visible universe at the range of the largest telescopes, the cosmic sphere could be packed with 10^{100} grains of sand — far more than the 10^{80} particles in the universe. Nor is the matter of the cosmos so tightly packed: each cubic centimeter contains on average only a few atoms.

NEW CONSTANTS FOR OLD

When we talk of fundamental constants, it is worth noting that things are not always as "fundamental" as we think they are. These fundamental constants are not only — as we remarked at the start — a measure of our ignorance; they also differ in quality in that some are "more fundamental than others," and frequently have a checkered history. They may fall from grace at any time and so be cast out of the circle of the chosen few, the "true" fundamental constants.

Before we go on to analyze a principle which seeks to explain real and accidental connections between the fundamentals of the physical world and our own existence, we need to have a clear understanding of the nature of the natural constants. In the closing section of this chapter, therefore, we will take a longer look at this; we can then proceed to examine the consequences of the anthropic principle for different fields of knowledge.

To take as an example the Hubble parameter, the reciprocal of which is a measure of the age of the universe, and which at the same time tells us at what speed the distant galaxies, in their recessional flight, are separating from us (and from each other): the value of this constant changed radically since it was first measured by Hubble in the 1920s — mostly in the direction of a smaller figure. This is not because the speed at which the universe is expanding has dramatically reduced since the twenties, but simply because astronomers have continuously refined their observation techniques and thus have frequently been compelled to revise the calibration of their cosmic yardsticks — such standard measures of the universe as binary stars, standard galaxies and luminous gas clouds.

How Constant are Fundamental Constants?

Alongside the question of the reliability of our measuring techniques in the area of the fundamental constants — central though this consideration is — there are a number of other questions to be asked in this connection:
— Why do fundamental constants only exist for physics and not, for example, for biology or geology?
— Why do some classical theories of physics contain no such constants at all (such as, for instance, classical mechanics)?
— Are the atomic constants, the velocity of light and Planck's constant, in some way "more fundamental" than other constants, for example those of thermodynamics or the theory of solid bodies? Examples here are thermal capacity or coefficients of elasticity for different materials.
— Are these atomic constants on a par with other data, like the masses of elementary particles and the constants of the four forces of nature; or are they more comparable to the factors used to convert kilometers to miles or centimeters to inches?
It is worth mentioning in passing that the systematic clarification of these (and other) questions was not begun until the 1970s, most recently and significantly by the French theoretician J. M. Levy-Leblond of the University of Paris. In 1977 he suggested that the constants should be divided among three groups:

Group A: Constants determining physical properties of certain individual objects, e.g. the masses of the elementary particles, their magnetic moments, radioactive decay times, and so on.

45

Group B: Constants that determine entire classes of physical phenomena. This group includes the constants of the four interactions, i.e., the four fine structure constants along with the charges on whole families of particles.

Group C: Universal constants such as the velocity of light and Planck's quantum, which occur throughout the edifice of theoretical physics and are not, as is the case of the propagation of light and quantum theory, limited to specific processes or individual theories — that is, constants that are not bound to individual objects (*Group A*) or particular forces (*Group B*).

The term "universal" applied to the constants of *Group C* merits a brief comment. All of these quantities should in fact be universal to be eligible for the name "constants" — that is, they should be independent of place and time in the cosmos. It is the task of experimental physics and observational astronomy to ascertain whether this is indeed the case.

In the well-known theories like Einstein's theory of gravitation, it is taken as read that the constants — the gravitational constant, for instance — are universal and completely constant. Whether Dirac's concept of a "time-dependent gravitational constant" can have validity — and thus be beyond the reach of current theories — will be examined shortly (see Chapters IV and VII).

The status of a fundamental constant and the group it is assigned to is very much dependent on the level of knowledge at any given time — as a glance at the history of science indicates. New findings may trace a particular constant back to more fundamental quantities — for example, the masses of atomic nuclei can now be calculated from our improved knowledge of the individual nuclear particles. Just the same fate overcame other "constant" properties of macroscopic bodies — their respective densities, compressibilities and thermal capacities (their capactiy to store heat), for example. At the turn of the century, these were all considered "fundamental" and thus counted among the fundamental constants.

Even the masses of the elementary particles — at least those of the proton, neutron, and mesons — may soon be derivable (and therefore "explainable") from still more fundamental units, the "quarks" (see Chapter III). Leptons — that is, neutrinos, electrons, muons, tau particles, etc. — are still held to be "fundamental"; they are assumed to have no further inner structure.

As a further example of the career of a constant, we can take the charge on the electron. This began as a property of one specific particle only, and thus belonged to *Group A*. But then it was discovered that, in the form of the electromagnetic fine structure constant, it characterizes the full range of electromagnetic phenomena, and as a result the elementary electric charge moved to *Group B*. The same course was followed by the velocity of light which, at the time of James Clerk Maxwell, that is from 1864, was also recognized as the speed of propagation of light waves, and thus belonged to *Group A*. But since the arrival of Einstein's twin theories of relativity (the general and the specific) and their concepts of space and time, it is clear that the velocity of light is basic to *all* processes in space and time, and even to the structure of space and time themselves, and that it thus plays a central role in all elementary processes whether or not these are electromagnetic likewise migrated to *Group C*.

When, in the discussion of Man as the yardstick of the physical world, we change physical laws solely by varying their constants rather than altering their form, we can do this in two ways:

— We may allow the possibility that the fundamental constants, and thus the laws of nature, are not universally valid; that is, that they may depend for their numerical value on their position in space and time;

— Or we may consider as a possibility that the laws of nature may remain universal, but have *other*, universally constant, constants.

In both cases, our aim is to establish how well the hypothetical changes made to the laws of nature match our observations, and above all to what extent they help or hinder the coming into being of an intelligent life form.

Chapter III
Fundamental Particles
With Charm

THE INSIDE OF MATTER

The road taken by the physics of the smallest building blocks of matter since the mid-seventies has led from nuclear particles, the building blocks of the atomic nucleus, to quarks, the building blocks of nuclear particles. In this chapter, as we consider these developments, we will use the anthropic principle to deduce some of the properties of matter at this most fundamental level that are necessary for the existence of life as we know it.

This subject — the elementary building blocks of matter — touches on a number of questions including philosophical ones, that are basic to natural science. In the last few years, the physics of elementary particles has experienced a flood of experimental results, and at the same time undergone a conceptual revolution comparable with that of the twenties, when quantum mechanics was developed. The award of the 1979 Nobel Prize for Physics to the high-energy physicists Steven Weinberg, Abdus Salam, and Sheldon Lee Glashow is just one sign of this.

This flood of newly-discovered particles and new theories upset some scientists' previously cherished beliefs. In their search for simplicity, some single and unchanging basic unit of matter, they found themselves faced instead with a confusing multiplicity. The assumption that the hundred-odd known elements were built from just three types of particles — the proton and neutron (in the atomic nucleus) plus the electron (in the shell of the atom) — satisfied their expectations for some years. But the road from these three particles to the quarks and leptons now held to be more fundamental — around four dozen new particles all told — was also marked by deep disillusionment that continually soured the natural euphoria at each new discovery.

Given that the high-energy experiments in particle accelerators such as CERN (Geneva, Switzerland), DESY (Hamburg, West Germany), SLAC (Stanford), or FERMILAB (Chicago) regularly produce end products that are more complex than those input, these doubts regarding the "fundamental" character of many of the nuclear building blocks are not hard to understand.

But the problems presented by the foundations of nature go further. They even have implications for science's concept of reality, which — since the arrival of quantum mechanics — is already unable to get by without including the relationship between the observer and that which is observed. "A phenomenon is only a phenomenon if it is an observed phenomenon," the American physicist John Archibald Wheeler (Austin, Texas) has said. This statement is certainly true as far as microphysics is concerned; how far it can be applied to the universe at large is still unclear. But it may well be that the existence of an intelligent life form, an observer of the cosmos, affects the reality of the universe.

All these questions are the subject of this chapter. But first, one particular consequence of the anthropic principle must be considered. This concerns one of the everyday characteristics of our universe: the stability of matter — that is, the fact that matter obviously exists over long periods and has been in existence, not just during recent times, but clearly for many billions of years.

The Stability of Matter and the Anthropic Principle

No human brain would even have been in a position to consider the relationship between life and the cosmos if matter did not exist in our present world and if it were not stable over periods of time sufficient for chemical and biological evolution to produce higher life forms. Humanity would never have come into existence, and probably the planets and the heavier elements would not have either. The existence of the Earth alone requires that matter must have been stable over a period of at least 4.5 billiion years. Besides this, all the chemical elements that served as raw material for the development of life on Earth must have originated far earlier, through nuclear fusion in the interior of stars, and the lightest elements, hydrogen and helium, right at the start in the Big Bang, the hot birth of the universe. This adds several million more years to the time required to accommodate the heavy elements; and to allow for light elements the total is between fifteen and twenty billion years — almost the entire lifetime of the universe.

All the stable elements must have remained truly "stable" over these enormous periods of time: otherwise mankind would never have looked out on the light of the stars. In order that a universe can bring forth observers, then, it must guarantee the conditions that are necessary for the stability of matter — conditions which must also still prevail today. Even physicists, however, were far from spotting that this presented anything in the way of a problem. It is taken too much for granted as part of our everyday existence that we are surrounded by more or less stable matter — rocks, water, air, refrigerators, nuclear power stations — for us to suppose this conceals a scientific poser. Anything so simple and obvious, one would think, must have a simple and obvious physical explanation.

The opposite is the case. "Some features of the physical world are so commonplace that they hardly seem to deserve comment. One of these is that ordinary matter, either in the form of atoms or in bulk, is held together with [electrical] Coulomb forces and yet is stable," as Elliot H. Lieb, one of the physicists to have been intensively concerned with the problem, wrote in 1976. By then, however, this irksome gap in our understanding of the cosmos had already been bridged, not least by Lieb himself.

The stability of matter, "this truly remarkable phenomenon" (Lieb), follows from the theory of quantum mechanics developed in the 1920s, which finally offered a comprehensive scheme for atoms built of nuclei containing protons and neutrons and atomic shells consisting of electrons. But for some years the problem was left in abeyance; and it was not until 1967 that two of America's leading scientists, Freeman J. Dyson and A.

50

Lenard, finally led "a heroic frontal attack on this difficult problem," as the Viennese theoretician, Walter Thirring, described it.

The "Coulomb forces" mentioned by Lieb are none other than the force of electrical attraction that two electric charges — one positive and one negative — exert on each other. (Two positive or two negative charges repel each other.) This force diminishes by the square of the distance between the charges; to this extent, the Coulomb forces act similarly to the force of gravity between two bodies. (Reports of a fifth force claiming deviations from this law at a scale of tens to hundreds of a meter are as yet inconclusive.)

Let us carry out a small, imaginary experiment. If we take two (opposite) charges, one in the left hand and one in the right, and let them go, they will attract each other more and more strongly the closer they approach each other: they will move towards one another until... until what? Until they collide? Can this really hold good if the negatively charged shell of each atom is to maintain a minimum separation from its positively charged nucleus? Even the nucleons of the nucleus itself, while they are tightly packed, can only be forced together up to a certain point. Wouldn't this situation lead to continual catastrophes, for instance, in the case of the most common atom in the universe, that of hydrogen, in which one proton and one electron circle each other? How could atoms ever exist in the expanded form we know? Embarrassment over this "awkward matter" in theoretical physics was expressed as long ago as 1915 by the British astrophysicist J. H. Jeans. Shortly before the advent of quantum mechanics he noted that "there would be a real difficulty in supposing that the law $1/r^2$ [stating that the attraction is inversely proportional to the square of the distance] held down to zero values of r [i.e. zero distance]. For the force between the two charges at zero distance would be infinite, we should have charges of opposite sign continually rushing together and, when once together, no force would be adequate to separate them.... Thus the matter in the universe would tend to shrink into nothing or to diminish indefinitely in size."

Why, then, do atoms not collapse in on themselves?

When the principles of quantum mechanics crystallized in the early

1920s — as a result of the work of Erwin Schrödinger, Wolfgang Pauli, Pascual Jordan, Neils Bohr, and others — the Austrian physicist Paul Ehrenfest was struck by the remarkable fact that every atom actually consists almost entirely — up to 99.99 percent, in fact — of empty space: "We take a piece of metal. Or a stone. When we think about it, we are astonished that this quantity of matter should occupy so large a volume. Admittedly, the molecules are packed tightly together, and likewise the atoms within each molecule — but why are the atoms themselves so big?" This size of each atom — for instance, that of hydrogen — is explained by the "uncertainty principle." The closer the orbit of the electron is to the proton, the higher its orbital velocity. From this can be derived a smallest possible radius for the electron's orbit: the minimum orbit available to the electron is 100,000 times as large as the diameter of the proton — which is why atoms are so suprisingly large.

If there are several protons and neutrons in the nucleus of the atom and several electrons in its shell, "peaceful coexistence" among these particles is regulated by a further law, known as the Pauli Exclusion Principle. This is actually a prohibition, stating what may not happen rather than what does: its effect is, in short, that no two particles may exist in the same state. Thus, two electrons may not occupy the same "room," but must make do with neighboring rooms. And when the ground floor is full, they must climb to the next floor. In the shell of the atom, the ground floor offers two rooms, the second floor 8, the third 18, and so on. Analogous "stories" regulate the arrangement of the particles in the nucleus. Dyson and Lenard's "heroic frontal attack" of 1967 proved that, *without* the Pauli principle, the force of electromagnetic attraction would mean that "not only individual atoms but matter in bulk would collapse into a condensed high-density phase. The assembly of any two macroscopic objects would release energy comparable to that of an atomic bomb."

Let us examine this Pauli principle. It was formulated by Wolfgang Pauli in 1935 and represents a firm statement regarding the most important characteristics of a particular group of particles including the electron and proton. It states that if there are several identical particles in the nucleus or the shell of an atom, then each must occupy a different state. Thus, for instance, no atom may contain two electrons which are equal in their position and their other quantum characteristics such as their spin.

The group of particles affected by the Pauli principle includes — not by chance — all the particles involved in the make-up of the atom: electrons (and their anti-particles, positrons), protons and neutrons as well as all atomic nuclei consisting of an odd number of particles; to these can be added neutrinos. The quality that unites them, their "spin" — a sort of internal angular momentum — goes by odd halves (1/2, 3/2, 5/2, etc. in appropriate units). In recognition of the work of Enrico Fermi, these are known as *fermions* or "Fermi particles." Particles with whole-number spin (0,1,2, etc.) are not subject to the Pauli principle: these include photons, mesons, and atomic nuclei with an even number of particles — the *bosons* ("Bose particles").

SPACE MUST BE THREE-DIMENSIONAL

The size of the atom, the distance between nucleus and shell, was the second essential property of matter to become comprehensible once the principles of quantum mechanics were established. In approaching the nucleus of the atom, the electrons are forced, so to speak, to take up an orbit that stays above a certain minimum level of energy. Instead of falling into the nucleus, the electron increases its orbital velocity, and this arrests its fall at a fixed minimum distance from the nucleus — a consequence of Heisenberg's uncertainty principle. It is thanks to the Pauli principle and the uncertainty principle, then, that matter has acquired the necessary size and stability!

Lenard and Dyson, in putting together their proof of the stability of matter, made use of an everyday quality of space: the fact that the physical world exists in three dimensions. We can see that this applies to the reality we know by the fact that we need just three figures to fix the location of any point — one for each of the three dimensions length, breadth, height. For a two-dimensional surface only two figures are required; and in a four-dimensional space four would be necessary.

There is a close link between the three-dimensional nature of space and the fact that the force between electric charges falls off by the square of the increasing distance between them. This seems simple and logical and is in any case taken for granted as a facet of everyday life; but to prove it required a tour de force of mathematics we can only regard with awe. Dyson and Lenard solved the problem in the sixties; and in 1975 Lieb and Thirring were able to simplify the proof considerably and improve it.

Let us return to the anthropic principle and establish the main points that are essential to the existence of the universe of an intelligent observer:

— There can be no abundance of matter with long-term stability without the Pauli principle and the uncertainty principle;
— both must be valid over cosmic periods of time;
— space must be three-dimensional.

In a different cosmos, with other spatial dimensions, matter might well be stable in a different way from that established for "our " universe. There would likely be quite different fundamental principles and natural laws in place of the uncertainty principle, the Pauli principle, and electromagnetic force. About this we can only speculate. Nor can we exclude the possibility that a cosmos in which the Pauli principle is not valid might contain quite different varieties of stable matter and thus might even have produced totally different kinds of intelligence.

THE LIFE EXPECTANCY OF THE PAULI PRINCIPLE

It is not enough for the Pauli exclusion principle to be established in our universe. It would also need to hold good everywhere — at least within our own galaxy — for as long as the evolution of life called for it to apply. As this process would also have to include the origin of inanimate matter, and the chemical elements would have to survive at least until our own time, it follows that the Pauli principle would have to be valid for the entire age of the universe — around fifteen to twenty billion years.

This presents a tricky problem for the experimenter: how is he to test whether the Pauli principle was indeed valid fifteen billion years ago? Or, to put it another way, how often is Pauli's principle violated in the present-day physical world? Might there be a small failure rate that would be evident in some minor deviation from the expected result? And would this be large enough to be measured? Two physicists at the University of California at Irvine, F. Reines and H. W. Sobel, designed an experiment to answer this question in 1974. Its purpose was to examine the "durability" of the Pauli principle for at least one concrete example: the inner atomic shell of a specific chemical element — in this case iodine. Their results confirmed "a lifetime for such a violation, per iodine atom, of [at least] 2×10^{27} seconds." This is ten billion times the age of the universe! How did Reines and Sobel reach this result, with its comforting implications for the stability of matter? Obviously, they could not perch in front of each individual iodine atom and wait to see how many billion years it lasted before degenerating.

Their solution was to approach it statistically. Instead of observing a single atom for a very long time, they examined a large number of atoms for a short time. The atoms they observed were the iodine atoms in a sodium iodide crystal of 1.3 kilograms. A crystal of this size contains some 10^{27} separate iodine atoms. Any violation of the Pauli principle would be evident from weak X-rays originating (in the absence of other well-known mechanisms) from electrons within the atomic shell "falling" into inner orbits. A minute amount of X-radiation was indeed found; but it was so weak that, when it was spread over the huge number of atoms in the crystal, the experimenters were able to predict an extraordinarily long life expectancy for the Pauli principle.

If the weak anthropic principle had been used to "predict" this result *a posteriori*, it could not have given such a precise figure, but could only have set a lower limit for the lifetime of the Pauli principle — perhaps ten or one hundred times the age of the universe — that would be long enough to ensure that even large accumulations of matter such as the mass of the sun could survive up to the present.

EINSTEIN'S DREAM AND THE UNITY OF NATURAL FORCES

The many anthropic relationships in the blueprint of the physical world demonstrate how precisely *homo sapiens* fits into its scheme. But they also hint at deeper, still undiscovered physical relationships, behind which it is fair to suspect there may lie a wider unity of nature — matter as a precondition for intelligence, and intelligence as a precondition for the recognition of matter and its laws.

The history of science is marked by two opposing movements: a tendency to multiplicity and a tendency to unity. The trend to ever more numerous and ever narrower specializations has led to sharp criticism of scientists, questioning their fondness for arcane experiments producing immense tables of figures and pillorying the researcher who "knows" more and more about less and less. Erwin Chargaff, in his book *Unbegreifliches Geheimnis* (Incomprehensible Secret), was not the only one to offer a sarcastic commentary on the loss of the wide view and an overall understanding brought about by this fragmentation of scientific knowledge. Carl Friedrich von Weizsäcker also concerned himself with this question, and attempted to define the trend through the number n of groups into which one might divide experienced physicists, so that no one in any one of the groups understood the specialty of anyone in the other groups: "One hundred years ago n was probably still unity: every good physicist understood all of physics. When I was young, I would have estimated $n = 5$. Today [1966] n is probably a moderately large two-figure number." And fifteen years later, one might add, it has probably exceeded 100.

However, Weizsäcker also believed that physics possesses "a greater real conceptual unity than ever before in its history" and believes it to be "an ultimate task awaiting physics to bring the field to full conceptual unity, and that this task, if mankind has not in the meantime destroyed itself physically or intellectually, will be completed sometime in the future."

Within the sub-atomic and astrophysical research of the last two decades, the drive for a unifying overview has frequently come up against obstacles in the form of surprising individual discoveries, usually the outcome of specific experiments which appeared to be in conflict with each other. Concepts, schemes and theories vied for a place on the pedestal of eternal truth, and most fell back in short order. (The science-historian Thomas S. Kuhn has produced an analysis of this succession of theories in his book). But despite this "allegro furioso" of science, the present day still provides confirmation of Weizsäcker's historic finding that science develops "from unity, through multiplicity, to unity": "In the beginning we have the unity of the original model. This is followed by a multiplicity of practical experiences, the understanding of which opens up the potential of the model — indeed, makes possible a planned development of the model. The insights provided by these experiences modify the model in their turn. This brings about a crisis for the original model; in this phase unity seems completely lost. But the final product is the unity of a new model which encompasses in its details the multiplicity of findings.... Heisenberg formulated for this the concept of the closed theory. But a theory is no longer the unity of a scheme embodying plurality, but the unity of a concept which prevails over the plurality."

Among the concepts that appear to have gained a foothold in the confusing and varied theory of elementary particles are some that sound more at home in our own humdrum world. Alongside terms like "quark" and "quantum chromodynamics" we find such concepts as "flavor dynamics" and particles available in various "colors" with or without added "charm" or "strangeness." These names from the new cuisine of theoretical physics may well mark a new "unity in plurality" — at least in our understanding of matter. This, however, is only the end result of a long, step-by-step process of to-and-fro between unity and plurality — a process that has led in at least five stages to the present synthesis, which is probably far from being the final one.

Synthesis One: In 1666 Sir Isaac Newton drew up his Theory of Gravitation by means of which he demonstrated that all motion can be explained using a single law — both the terrestrial motions examined by Galileo and the astronomical motions already described by Kepler's laws.

Synthesis Two: James Clerk Maxwell showed in 1864 that the seemingly disparate phenomena of magnetism, electricity and even light could be represented by a single theory, and that all of these — previously examined separately by Faraday, Oersted, Ohm, Maxwell himself and others — could be based around a new concept, that of electromagnetic fields.

Synthesis Three: Albert Einstein's Special Theory of Relativity of 1905 united the classical mechanics of Newton and Maxwell's electromagnetic field theory.

Synthesis Four: In 1915 Albert Einstein set up a new theory, the General Theory of Relativity. This linked its predecessor, the Special Theory, with gravitation to provide a unifying concept, that of "curved space-time."

We are now in a period of scientific trial and error whose extent of over fifty years gives only a faint indication of the difficulty experienced in "marrying together" the fundamental forces of nature. The beginning of this period saw the development of quantum mechanics which provided an understanding of the motion of bodies on the nuclear scale. In the fifties, the neutrino and the pion — the first of the mesons — were discovered, the theory of the weak force and the first proposals regarding quarks were introduced, and Maxwell's theory was expanded to quantum electrodynamics. This abundance of unconnected threads of empirical knowledge presents the task of unifying all the new theories within the framework of a great, all-encompassing theory.

There has been no shortage of attempts. In the thirties, Hermann Weyl constructed a theory of gravitation and electromagnetism. Einstein devoted his last twenty years to similar efforts, and Werner Heisenberg thought that his "universal formula" had solved the riddle of the unity of natural forces, in principle at least. But none of these proposals was quite acceptable. The strongest resistance to a unified treatment is at present offered by quantum mechanics and gravitation.

It was not until the sixties that there was progress in such a form that scientists were hardly aware of it at first:

Synthesis Five: The Harvard physicist Steven Weinberg (then at MIT), with Abdus Salam (Trieste, Italy) and John C. Ward (New Zealand), succeeded at the end of the sixties in finding a link between the weak and electromagnetic interactions which they showed to be different manifestations of a single "electro-weak" force. In 1983, this theory was

58

gloriously confirmed by experiments in CERN in which the so-called w- and z-bosons were found — particles predicted by the electro-weak theory.

Since then we may have seen the advent of

Synthesis Six (?) : "Supergravitation" has been under examination since about 1971 in a cooperative venture involving a number of physicists in the USA, Western Europe, and the USSR, particularly the American physicists Daniel Freedman, Stanley Deser, Bruno Zumino and his European colleagues Peter van Nieuwenhenzen and Ferrara. They aim high: their goal is to unify all four natural forces and trace them back to a common force. The question mark against Synthesis Six indicates that it is still an open contest as the theories have yet to be fully developed and different ideas are still competing with one another. However, supergravitation predicts a new particle with a spin value of 3/2: the "gravitino."

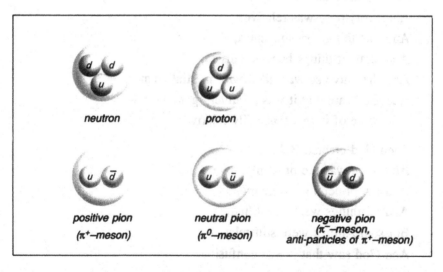

Every hadron, such as the neutron or proton, is composed of three quarks, while the pion and the other mesons consist of a quark-antiquark pair. The characteristics of the component quarks, added together, give the characteristics of the "whole" particle. (The bars over some symbols indicate antiquarks which have the opposite electrical charge to the matching quark.)

Unified Field Theory
Tim Joseph

In the beginning there was Aristotle,
And objects at rest tended to remain at rest,
And objects in motion tended to come to rest,
And soon everything was at rest,
And God saw that it was boring.

Then God created Newton,
And objects at rest tended to remain at rest,
And objects in motion tended to remain in motion,
And energy was conserved and momentum was conserved
    and matter was conserved,
And God saw that it was conservative.

Then God created Einstein,
And everything was relative,
And fast things became short,
And straight things became curved,
And the universe was filled with inertial frames,
And God saw that it was relatively general, but
    some of it was especially relative.

Then God created Bohr,
And there was the principle,
And the principle was quantum,
And all things were quantified,
But some things were still relative,
And God saw that it was confusing.

Then God was going to create Furgeson,
And Furgeson would have unified,
And he would have fielded a theory,
And all would have been one,
And it was the seventh day,

One theory that supergravitation would subsume as a component of the new explanation of nature is that of quarks. The basic concepts of quark theory were published in 1964 by Murray Gell-Mann of the California Institute of Technology who received the Nobel Prize in 1969 for his work.

QUARKS AND LEPTONS AT PLAY

When the idea of quarks was introduced, the particle menagerie was not as yet particularly full. The new idea was suggested independently in 1963 by two physicists at the California Institute of Technology, Murray Gell-Mann and George Zweig. Their aim was not a modest one: quarks were finally to furnish something men had been trying to find for at least 2000 years — the most basic indivisible components of matter.

The original version of the idea involved only three quarks, named "up," "down" and "strange." These were enough to build up the two most important stable hadrons, the nucleons proton and neutron. All atomic nuclei are composed from these two nucleons. In the framework of the quark theory, proton and neutron consist of three quarks each, the proton representing the combination up-up-down and the neutron the combination up-down-down. Under the simple rules for the assembly of quarks (and their corresponding anti-particles, the anti-quarks) into larger particles, these were the only two hadrons that could be built up in this way.

Each quark possesses only a fraction of the elementary electric charge, the up quark having two-thirds positive and the down quark a third negative charge. The electrical charge carried by a proton $(+1)$ thus derives quite simply from the sum of the partial charges of its quark components: 2/3 (up) plus 2/3 (up) minus 1/3 (down), giving the proton's full unit positive elementary charge. The electrically neutral neutron derives its zero charge as the sum of 2/3 (up) minus 1/3 (down) minus 1/3 (down). Unlike the nucleons, the particles in the other group, the mesons, are composed of only two quarks.

And what of the third quark? The strange quark was first brought in to explain the relatively long life of some heavy hadrons discovered in the fifties. Heavy hadrons usually decay via the strong interaction, which operates very quickly. One might say that the strong nuclear force is as strong as it is because it is so quick. A strongly decaying hadron generally exists for only 10^{-23} seconds before it breaks up into smaller components.

But one member of the hadron family, the lambda particle, did not follow this rule. It has a lifetime of 10^{-10} seconds, exceeding the typical strong decay time by a factor of ten thousand billion. It was clear that the lambda was not decaying via the strong force, but with the help of another force, the weak interaction, and with a decay time characteristic of the weak force. This eccentric behavior was readily explained with the quark model: the lambda contains "strangeness" in its quark make-up, which must be retained unchanged in one of the fission products when it reacts with the strong force. The lambda, however, is the lightest of the strange hadrons, and it cannot pass on its quota of strangeness to smaller particles after strong-force decay. For this reason it has to wait, so to speak, for the weak force whose considerably longer decay time allows the lambda to decay after all — since in weak-force decay the strange quark can convert into one up and one down quark. On the basis of processes ruled by the strong interaction, this decay chain would not be possible, so it uses the route via the weak force which takes longer.

In late autumn 1974, however, the carefree world of the three quarks abruptly collapsed: a "November Revolution," as the event came to be called, seemed to throw this near-holy trinity to the winds. Independently at two American laboratories — Brookhaven (New York) and Stanford (California) — a new particle was discovered; Samuel C. C. Teng at Brookhaven christened it "J," while Burton D. Richter of Stanford called it "Psi." One thing soon became clear: the J/Psi particle shattered the three-quark scheme.

The November Revolution soon produced Christmas 1974's gift to particle physics, since it made it obvious that there must be a fourth quark. The "charmed" quark — as quark No. 4 came to be called — helped explain the collision events that had remained inexplicable up to then, just as strangeness had explained the behavior of the lambda particle. The long life of the new particle could be understood on the basis of charm, in much the same way as strangeness had been able to account for the abnormal survival of the lambda.

The subsequent exact analysis of the experiments gave rise to a curious situation. The new J/Psi particle, a meson, was clearly composed of two charmed quarks, one with charm and one with "anti-charm." As they are each other's anti-particles, the pair have exactly opposite charm characteristics, a positive helping of charm for the one and a negative helping for the other. Added together, they cancel each other out. Seen from the outside, therefore, the J/Psi particle seems not to possess any charm at all — it only carries "hidden" charm.

According to today's quark model, it is believed that all mesons, like the J/Psi meson, are built from one quark and one anti-quark. It is only if just *one* of the two — or at least one of the three in the case of baryons — is a charmed quark that the charm becomes visible and can be seen from outside the particle: "naked" charm!

FLAVORS AND COLORS OF THE QUARK

With the four types of quark introduced up to 1975 — up, down, strange and charmed, the four quark "flavors," as scientists soon christened them — physicists were able to build up new particles in theory and predict their existence. Some of these predicted particles were discovered in ingenious particle accelerator experiments, and the long-contested idea of quarks seemed to have come through to victory. A zoo filled to overflowing with around 300 "elementary" particles had at a stroke been reduced to four new, even more elementary particles.

But the quarks' behavior, even after the expansion of their number into a "Gang of Four," soon began to present new problems. Like a child experimenting with his new building set, researchers began trying to build up familiar particles, hadrons and mesons, from the quarks they had discovered. The first moves in this direction were made in 1964, and it soon became clear that nature was unwilling to come up with all the varieties of particle that could be made from the pieces of the Quark Puzzle. Not all combinations of different quarks found counterparts among the particles of the real world — or else they were yet to be discovered.

And there was a further complication. A theoretical "selection principle" demanded that no two identical quarks could be united in one particle. This principle is similar to the one ruling the atomic nucleus, where no two nuclear particles may occupy the same state — requirement of the Pauli principle. In a neutron, however, two down quarks are housed with an up quark; and in the proton the down quark faces two up quarks. To save the quark concept, researchers introduced the postulate that each type occurs in three different versions. These versions were called "colors" and named red, blue, and green — so that there would now be three down quarks, for example: a red, a blue and a green down quark.

These different color varieties had an additional significance: they were to combine according to optical rules. If the colors red, blue and green are added together, the result is white light. In an analogous way, each particle appears externally neutral — that is, white — in the quark model, and this simple rule determines the colors of the quarks. In the neutron, each of the three quarks takes a different color. In this way, the occurrence of two completely identical quarks within one particle is avoided; and the sum of the colors produces the external effect of the neutral white. The two down quarks in the neutron are then distinguished by their color. This made it easy to find a name for the new theory: *quantum chromodynamics*. The name is analogous to quantum electrodynamics, which instead of describing the forces prevailing between the colored quarks describes those between the electrical charges.

Before long, it must be said, quantum chromodynamics began stocking a sizeable quark zoo — a disappointment for those who thought the search for elementary components of nature would lead to increasing simplicity. As each quark could occur in three different colors, there were by then — including all the anti-quarks — a total of 24 quarks. However, every known hadron could now be explained without problem as the combination of different quarks. In all the baryons each of the three quarks is of a different color; in the case of the mesons, which consist of only two quarks, the color of one quark is cancelled out by the anti-color of its complementary anti-quark. The quark model, with each quark taking three color varieties, seemed to function adequately.

Up to now, quark theory has not been able to explain why the quarks are never observed on their own. So far, it has only been possible to "sense" their existence within hadrons and mesons. However, particle physicists remain optimistic on this point.

THE GENERATION GAME: WHY DOES NATURE REPEAT ITSELF?

In research into the foundations of the natural world, the principles of symmetry play almost as dominant a role as the elementary particles themselves. Which of the two is the more fundamental is still a matter for argument. But there is no question that scientists see symmetrical principles as a tool that will help them to come to grips with the rapid proliferation of quarks as much as of the zoo particles.

Symmetry as an organizational principle has proved increasingly useful in the history of physics as a guide to better understanding. In the case of the symmetry between quarks and leptons, the story in fact goes back to the thirties and to the explanation of radioactivity which dates from that period.

Soon after the discovery of the neutron, it was observed that the new particle was confusingly similar to the proton. The electrically neutral neutron looks very much like a proton with its positive charge removed. There is also the question of the nuclear reaction that converts a neutron into a proton — what is known as beta decay: in this reaction a proton and an electron (or beta particle) are produced. Since electrons are not attracted by the nuclear force that binds protons and neutrons together in the nucleus of the atom, the beta particle leaves the nucleus, which is thus one neutron poorer and one proton richer after the conversion. Beta decay is one of the mechanisms of radioactivity.

In beta decay a further particle is liberated alongside the electron: the *neutrino* (see also Chapter IV). This is a truly invisible building block of matter, which is not only immune to the strong nuclear force but also to electromagnetic force, and only interacts with other particles via the weak nuclear force. It is able to pass through the Earth without hindrance, and even a solid block of lead of thickness equal to the distance from the Sun to Alpha Centauri (4.3 light-years) could not stop it. Just as the neutron is the chargeless counterpart to the proton, the neutrino takes the role among leptons as the chargeless counterpart to the electron.

Thus, as long ago as the thirties, scientists appeared to have found a perfect symmetry between the hadron pair proton and neutron, and the lepton pair electron and neutrino (strictly, electron and electron neutrino). The hadrons react to the strong nuclear force, the leptons do not. Another reason to see these four particles as a complete set is that the sum total of the electric charges on the four particles is exactly zero: electron and neutrino have one negative charge between them, which is cancelled out by the positive charge of proton and neutron.

Today, this symmetry has yielded to a similar, but even more fundamental one between leptons and quarks; and instead of the several hundred kinds of hadrons, the scheme can be built up from relatively few quarks. The rule now is: for every lepton pair there is a corresponding quark pair. And each set of two corresponding pairs has been classified as a "family" or "generation" of particles:

— The up and down quarks, the electron and its neutrino form the first generation;
— The charmed and strange quarks along with the muon and its matching neutrino (the muon neutrino) for the second generation.

At this point, in 1974, the world of quarks seemed to have been put back in order: the symmetrical quartet provided a balanced and satisfying pattern. The division into generations, too (which can be thought of as an analogy to the arrangement of chemical elements in the periodic table), proved successful; for all the first-generation processes within nature there were parallels in the second generation. One aspect that remains unexplained up to the present day is why the second generation exists at all. "Why nature 'repeats' itself," mused the British researcher Frank Close in 1979, "physicists do not yet know. The familiar world about us appears to survive quite well with only one pair of quarks (up and down) and one pair of leptons (electrons and neutrino). These particles alone are sufficient to explain the structure of the atom and the interaction between different atoms — the basis of chemistry. But the existence of the second generation of particles seems to be an extravagance, and does not appear to affect our everyday experience in any significant way. We do not understand why the second generation is there at all."

And that, we might add, goes for the *third* generation as well! For the discovery of strangeness and charm was not the last surprise in store in the world of particles. Scientists' intuitive feeling, however, is that, although these high-order generations of particles play little part in the processes of the present (cold) universe, they must have taken a more central role in the hot environment of the very early phases of the Big Bang when matter was being created.

Key Questions on Matter (1): What Are Nature's Fundamentals?

The breakthrough into the third generation of elementary particles was achieved by Martin Perl at California's Stanford accelerator in 1975 when he discovered a new lepton, the "tau" particle. A matching tau neutrino was also discovered. In view of the symmetry between lepton and quark pairs that had held good up to then, a hunt for the corresponding third pair of quarks naturally got under way. The names held in readiness for the new arrivals were "bottom" and "top" — or, alternatively, "beauty" and "truth" (but either way, "b" and "t" for short).

The christening was soon followed by a birth. On May 1, 1977, Leon M. Lederman of the Brookhaven National Laboratory on Long Island announced his latest find: a particle whose construction matched that of the J/Psi meson. Like the latter, it was composed of a quark and the corresponding anti-quark; however, this time there was no charm, but instead the predicted bottom quark. The new-found "upsilon" particle, then, was built from a quark-antiquark pair of bottom quarks, and was the first piece of matter to be exclusively made up of components from the third generation.

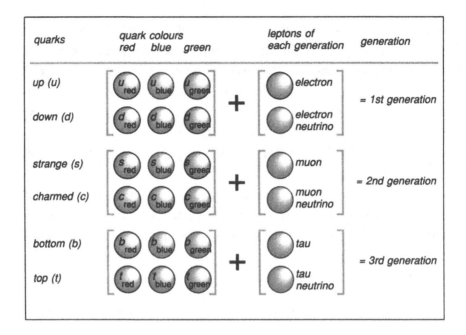

quarks	quark colours red blue green			leptons of each generation	generation
up (u)	u red	u blue	u green	electron	= 1st generation
down (d)	d red	d blue	d green	electron neutrino	
strange (s)	s red	s blue	s green	muon	= 2nd generation
charmed (c)	c red	c blue	c green	muon neutrino	
bottom (b)	b red	b blue	b green	tau	= 3rd generation
top (t)	t red	t blue	t green	tau neutrino	

Quark families:

Quarks can be classified in pairs, each pair combining with a corresponding pair of leptons to form a particle generation. Each generation is thus composed of

— *two quarks, each of which can take one of three colors: red, blue or green - a total of six;*

— *two leptons, one of them the neutrino of that generation;*

— *and a full set of corresponding anti-particles - a further eight.*

There are thus 16 members of each particle generation: 12 quarks and 4 leptons. All stable particles are composed exclusively of elementary particles of the first generation; generations 2 and 3 only appear at high energies. It is not yet known whether the three families shown here represent the full complement of nature's "quark zoo," nor whether there may be deeper cross-connections between these three generations.

Germany's DESY accelerator in Hamburg quickly found more upsilons; but the third generation was still incomplete. The total so far was six leptons and five quarks (not counting the color variations).

What was still missing was the top quark. In spite of great efforts, not least by the physicists in Hamburg, it has remained elusive although some indications were claimed to have been seen in 1983 in CERN experiments, though inconclusive.

This runaway proliferation of quarks and leptons, meanwhile, has made some scientists rather suspicious of the entire quark concept. For each generation of elementary particles there are now two quarks, each in three color varieties, making six quarks per generation. Added to this are two leptons per generation; augmented by the corresponding number of anti-particles, this amounts to 16 particles per generation, a total for the three generations of 48 elementary particles — threatening to become as large a number as those that had been left behind, and already too large for many who had hoped that the exploration of the fundamental level of particles would make for simplicity in numbers as well as in everything else. For many, any number over ten no longer smacks of anything truly elementary. Another, a weakness in the theory is that it has yet to come up with a reliable method of predicting the mass of a quark.

The worry is that the search for a scheme that is readily accessible to human understanding and at the same time provides mathematical simplicity may be leading researchers down a false trail.

One unexpected finding at least is the fact that quarks only exist "hidden" in the interiors of particles. Up to now no free quark has been observed. But this coyness, unique in nature, may have a positive side. "If quarks really are true 'elementary' particles," suggested the journal *New Scientist*, "then we might expect them to display some unique characteristics as they interact through the fundamental forces in nature."

Inside a proton, for instance, quarks appear to move around quite uninhibitedly — rather like marbles in a bag: but these marbles can never jump out of their bag. The force between quarks, transferred via the whimsically-named "gluons," is clearly constructed in such a way that it virtually disappears when the quarks are at close quarters inside the proton, but sharply increases when they move apart. It is as if the quarks were joined together by loose rubber bands, which are slack at short distances but stretched tight at longer ones.

The West German electron synchrotron (DESY = Deutsches Elektronen Synchrotron) in Hamburg-Bahrenfeld. (Photo: DESY)

Even the assumption that there may be a final, lowest level of simplicity in the description of nature in the first place may be an erroneous one. If it should be positively confirmed that nature's foundations rest on the pillars of 48 truly elementary particles, the scientific world will of course accept this with equanimity. But if proved otherwise, the search will start for other qualitites or principles that could stand in place of particles as something more elementary, even "the most elementary"; and these in their turn may well prove to be no more than a further intermediate step on the endless road into the interior of matter.

The quark concept, then, has met its share of critical questioning. We will deal here with just one of these questions: what is *inside* the quark?

Can we go even deeper inside the proton and find further levels of activity? How are quarks constructed? The "deep inelastic" probes into the interior of the proton with high-energy electrons have provided indications that quarks show minute deformations.

Photographic record of a particle collision. (Photo: DESY)

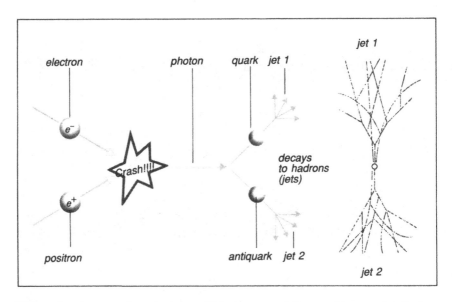

When electrons and positrons collide in a particle accelerator, the fission products record their tracks of photographic plates. The two jets on the right are the result of the subsequent decay of a quark and anti-quark.

These variations are explained by the emission of a "gluon": gluons, according to theory, are the mediators of the force connecting individual quarks. It is the gluons, in fact, that lock the quarks in their hadron prison. To understand the role of gluons we need only to consider electric charges: if an electron is accelerated, it radiates, emitting particles of light (photons). A quark in motion radiates in just the same way: it emits gluons. Gluons are thus the mediators of the strong nuclear force, just as photons are the mediators of electromagnetic force. Gluons were first observed in 1980 at the DESY research center in Hamburg — but not, of course, directly. Since free quarks are not directly observable, the existence of their gluons must also be indirectly inferred.

The "proof" of the existence of this "atomic glue" was only furnished by a complex chain of indirect arguments and interpretations. The more closely we probe into quarks with high-energy shots, the more we see

72

of them — witness the recently observed gluon deformations. If the current interpretations of the "flaw" hold up, it is at present unlikely that any more fundamental building blocks will be needed. However, it is interesting to note that particle physicists have already set the stage for the next level down: the "sub-quarks," supposed constituents of quarks.

KEY QUESTIONS ON MATTER (2):

ARE WE DROWNING IN A SEA OF PARTICLES?

At this point the consumer of scientific knowledge, a little nervous perhaps at the helter-skelter discovery of more and more particles, will be anxiously wondering where this headlong dash into the foundations of matter will end. The technology of the particle accelerator is being developed further and further to ever-higher energies; the Stanford Linear Accelerator and the LEP accelerator at Geneva both commenced operation in 1989. They work at energies above 100 GeV — a ring of 27 kilometers in circumference — so they may see the top-quark, or another fundamental particle thought to be responsible for the creation of mass, the so called Higgs particle. But each fresh enlargement of the particle zoo is likely to present scientists with a longer list of embarrassing questions and at least partially demolish the theoretical edifices built up with such painstaking toil. And in this research the *nondiscovery* of a particle can be as much of a surprise as a new particle's unexpected *discovery*; while a major problem lies in the fact that the masses of quarks themselves cannot really be predicted: in principle, quarks may not have any mass at all.

Is there some magic to the figure 6? We now have six types of quarks and six leptons: is that the end of the story? Or are there more quarks to be discovered — with greater energies, will we find further generations of particles? Why only six — why not eight or twenty-four? After the third generation, is there a fourth, a fifth...? And if the zoo of new elementary particles were to grow like the hadron zoo before it, what price then the belief that quarks really do represent the most fundamental level — the level at which we are left with only the final "indivisible" building blocks?

73

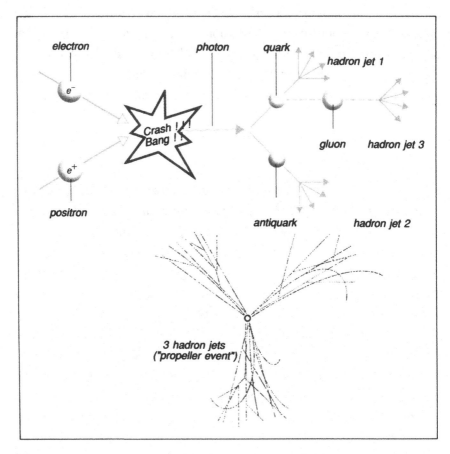

Before decaying into a jet of hadrons (e.g. Jet 1 above), a quark may emit a gluon: when this occurs, a third jet (Jet 3) will be observed, as the product of gluon decay. If the gluon itself is low in energy, jets 1 and 3 will combine; but at higher gluon energies the directions of the two jets diverge markedly, producing what is called a "propeller event" - an event recorded in DESY's TASSO research facility.

As long as we know as little as we do of the still mysterious quarks, these remain fair questions. This historical perspective shows that an overly quick growth in the number of "elementary particles" can justly be taken as an indication that these are not the ultimate components of material existence sought for so long. In the 19th century chemists traced the host of chemical compounds back to only three dozen chemical elements at first, but their number rose to over 100 in our own century. Then, in the thirties, a new simplification was introduced. The place of the chemical elements was abruptly usurped by three atomic particles: proton, neutron and electron. By this spectacular innovation all the chemical elements could be built up from these three particles, and arranged in matching groups of eight — the periodic table. As a result, after World War II, the first mesons were discovered: pions, kaons (or k-mesons), lambda particles... Soon there were too many.

Finally, 1963 saw the arrival of the quark concept and the possibility of a new simplification. Would three quarks now take the place of hundreds of hadrons? But now we are already hunting the sixth quark. And this means that, taken together with the three color varieties (red, blue, and green), experimenters are hoping to find quarks 16 to 18 — and this is without counting the anti-particles. However, the advent of quarks has for the first time brought a radically new aspect into the physics of matter — the fact that these particles can only exist in combination with others of their own kind. This makes quarks something special. It is a signal that we may now indeed have reached a decisive level of fundamentality.

But although quarks are not "free," and cannot be examined on their own, this has not impeded research into them. Another special characteristic of quarks is that, within their hadron "bags," they can move around almost as if they were free particles, especially when — at high energies — they come close together. This is their one measure of freedom in their particle prison. Is this the point at which nature has chosen to erect a wall to make it impossible, or at least difficult, to spy out the ultimate secrets? The discoverer of the upsilon meson, Leon Lederman of the Brookhaven Laboratory, risked an optimistic comment in 1978: "And who is to say that physicists will never build an ultra-powerful accelerator that could overcome the confining force and liberate the quark?"

Despite the impressive record of success achieved by the quark model and quantum chromodynamics, some scientists have been left with misgivings. These arise from concern as to what may be properly regarded as "fundamental," and may be inspired by visions of a future "Unified Field Theory" or "Grand Unification Theory." The encouraging process in this area is tempered by doubts as to whether the quark concept, while being retained for the moment in its obvious usefulness, should not now be stripped of its claim to define a "fundamental" level of physics. For quantum chromodynamics could be equally deserving of the mocking remark once made by Wolfgang Pauli with regard to the "universal formula" of Werner Heisenberg: it was, he said, "like a picture frame for a Titian masterpiece — all we need is the picture."

But Heisenberg, too, ciriticized modern elementary particle physics and the quark model. Instead of particles, Heisenberg preferred to see symmetries at the fundamental level of nature. He pointed out that with the elementary particles science had for the first time reached a level where the energy required to split the particles was of the same order of magnitude as that contained in the masses of the decay products. When elementary particles collide, "the fission products at these high energies are in no way more elementary than the particles which originally collided." Heisenberg and his school stressed, as Heinrich Saller noted that "the concept of the particle should be supplanted by that of symmetry." Saller, a researcher at Munich's Max Planck Institute for Physics, of which Heisenberg was director until 1970, also speculated whether quarks might not sooner or later suffer the fate of the ether, that hypothetical medium thought to have filled space and through which the Earth was believed to move on its cosmic travels. At the beginning of this century, Albert Einstein was able to "kill off" the ether in light of his Special Theory of Relativity as an unobservable and therefore scientifically useless ballast.

Nearly all physicists share an unshakeable belief in the simplicity and unity of nature, even though this is no more than a methodical principle used in the formulation of a new theory. In view of this, "it was only to be expected, when the proliferation of 'turrets' of flavor and 'towers' of color got out of hand, that misgivings at the rococo nature of the quark would arise" (Saller). Heisenberg's advocacy of symmetry is reminiscent of ideas put forward by Plato and Pythagoras 2500 years ago. Pythagoras insisted that all order could ultimately be understood by numbers and their harmonies. And Plato provided an explicit example of this in identifying the fundamental symmetries as the elements of nature. Translated into modern language, we would say that we should not pay too much attention to the final count of quarks and particle generations we need for a complete picture of matter, but should look for the simplicity of nature in the symmetry of groupings of the appropriate theory. One of the originators of the quark model, Murray Gell-Mann, now appears to share Heisenberg's viewpoint: in a lecture at the University of Munich in 1979 he said: "Simplicity is not an economy of particles but an economy of principles!"

It must remain an open question at the moment, then, whether the generation game of the elementary particles may not point to a closer link between quarks and leptons. One hint of this may be given by the number of fundamental constants any force needs in order to define it. The theory of electrons and photons, quantum electrodynamics, requires only two constants to define the force of repulsion between two electrons: mass and charge. Quantum chromodynamics, on the other hand, need no fewer than *seventeen* constants — sixteen masses and a coupling constant — to cover the details of the strong nuclear force. "This is clearly too large a number to be truly fundamental," commented Sheldon Glashow of Harvard University, winner of the 1979 Nobel Prize for Physics. There are other unsolved riddles as well; for instance, the quark model does not help explain why the electrical charge in the electron is exactly as large as, and opposite in sign to, that on the proton.

The aim of the "grand unification theories" is to find a way of solving these problems — and their originators are very much in sympathy with

Heisenberg's point of view. A basic element in these theories is the fact of the particle generations themselves. Beyond this, their proposers work on the assumption that matter will also demonstrate greater symmetry at higher energies, or higher temperatures. That is to say, the higher the energy, the more alike are the elementary charges, both in their strength and in the nature of their physical interactions. The way in which the symmetry of a system may increase as the energy level rises can be seen by comparing liquid and crystal. In a liquid all directions are as a rule equal in status: the liquid is "isotropic," corresponding to a maximum *spatial* symmetry. (Besides spatial symmetry, there are also symmetries in time.) In its cooled state, the liquid crystallizes — to ice, for example — and the crystals show preferred directions. The crystalline body is thus anisotropic, and construction therefore has a lower spatial symmetry.

THE SIXTH, SUPER-WEAK, FORCE OF NATURE

The unified field theories assume that at extremely high energies, way above 10^{15} GeV, such as must have prevailed in the cosmos for a brief moment in the Big Bang, all forces are equally strong and all charges alike, so that they could be defined by a single fundamental constant. It was only as the universe cooled a little that this unified original force "froze out" into three different, asymmetrical forms — that is, into three forces of different strengths. (The fourth force, gravity, is not included in this concept.) With the cooling and expansion of the universe, they went their different ways: the strong nuclear force grew considerably stronger, the weak nuclear force strengthened only slightly as the cosmos developed, and the electromagnetic interaction became weaker. Every version of the theory also predicts that the strong nuclear force will end up markedly stronger than the electromagnetic force.

The differences between the natural forces in the present, cool cosmos were essential for the development of structures — from the basic forms of matter to the evolution of biological systems. As long as all three forces remained equally strong, all particles would react to one another equally rapidly and equally strongly, so that no structures would be able to develop. As a result, all matter would consist of a fairly uniform, homogenous but highly energetic "soup," in which any "lumpiness" would be smoothed out again at once. Only after the nuclear forces had achieved dominance over electromagnetic forces could atomic nuclei develop. And only once the electromagnetic forces strongly outweighed gravitational forces in the microcosm could chemistry, and consequently the chemical evolution of atoms and molecules, come about.

Some of the processes of the hot Big Bang have been simulated in particle accelerators. However, it is only inside such accelerators and in cosmic radiation of space that it has been possible to observe the particles of the second and third generations produced in the first split-second of the young universe. It is interesting that the respective masses of quarks and leptons become more alike in the higher generations:

First Generation   **Second Generation**   **Third Generation**

$$\frac{d\text{-quark mass}}{\text{electron mass}} = 20, \quad \frac{s\text{-quark mass}}{\text{muon mass}} = 5, \quad \frac{b\text{-quark mass}}{\text{tau mass}} = 2,$$

but it seems one cannot draw any wide-ranging conclusions from this.

In this totally symmetrical birth of the universe (as far as forces are concerned), a further and otherwise improbable process would occur, which would convert baryons into leptons, and leptons into quarks. These events, like any other, must of course be mediated by a specific interaction, in this case an additional, sixth force of nature (beyond the fifth force sought for in the gravitational interaction). Today it could hardly play any part in the behavior of matter, since it would be weaker than the weak nuclear force — already very weak — by a further factor of 10^{29}. For this reason it has been named the "super-weak " force.

The prediction of a sixth interaction in nature is probably the most exciting result to date of the unified field theories. As this sixth force can bring about the conversion of quarks into leptons, it can in particular trigger the decay of protons. Under previous theories, the proton was treated as absolutely stable, a particle with infinitely great life expectancy. The decay of a proton, if one were found in experiments today, would of course be only the very weakest shadow of a process which was a common occurrence in the hottest and earliest phase of a Big Bang

79

with symmetrical interactions. And, indeed, calculations using standard "Grand Unified Theories" did predict a vanishingly small, but finite, positive probability of decay for the proton, corresponding to an average lifetime of 10^{28} up to 10^{31} years. This timespan is many times longer than the age of the universe, around twenty billion or 2×10^{10} years; so in terrestrial terms the proton would be everlasting.

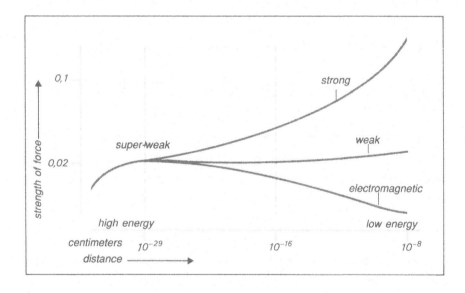

The "Grand Unified Field Theory" predicts that at short distances, corresponding to high energies, the strengths of the fundamental forces will approach the same value. This would mean that in the Big Bang, when the cosmos was still very hot and in a highly energetic phase, all natural forces would be more or less equal in strength. As the cosmos expanded and cooled they would then have followed separate developments. This difference between the strengths of the natural forces was essential for the origin of life.

Nevertheless, intensive attempts are being made to measure proton decay. Any life expectancy significantly shorter than the predicted one can quickly be excluded: if protons were to decay after "only," say, 10^{16} years — a period roughly a million times the age of the universe — every human being of normal weight would be radioactive. Each of us would show a measurable activity of 3 micro-curies, comparable with the radioactivity involved in medical examinations of the thyroid gland. In the mid-fifties, Frederick Reines, Clyde Cowan, and Maurice Goldhaber of the Brookhaven National Laboratory established a minimum proton lifetime of 10^{22} years; and in 1979 Reines and his colleagues were able to raise this value to at least 10^{30} years. Recent proton decay experiments claim about five times this value.

In spite of this fantastically small anticipated decay rate for the proton, there is now a possibility, not only of arriving at a lower limit for its life expectancy, but of measuring the value directly. There is a method for this: first collect as much water as you can — ordinary water as far as possible — and try observing all the protons at the same time. One ton of water contains about 10^{29} individual protons, so that this ton of water will provide enough for one proton to decay roughly every ten years; by amassing a large amount of water the number of decay events in a given time can be brought up to a level useful to the experimenter. Experiments in this direction have been performed since the early eighties at several places: in deep underground mines in the US, Japan and India, but also in the Frejus tunnel between France and Italy. At present (1989) they preclude a lifetime for the proton of less than 4×10^{32} years for the favoured decay mode, where the proton disintegrates into a positron and a (neutral) pion.

Protons are a vital necessity for the existence of matter. The validity of the Pauli principle, the fundamental anchor of the stability of matter, also depends on a high life expectancy for individual protons and neutrons. The anthropic principle which seeks to specify the conditions indispensable to the existence of life, demands that this life expectancy must be very great and amount to at least a billion times the age of the universe. At the same time it must postulate that the super-weak force no longer has any effect in our universe, and that it thus really is *super*-weak.

"Grand Unification" represents at least a first step in the realization of Einstein's dream of understanding all the forces of nature as expressions of a single (symmetry) principle. As yet, gravity has still to find a proper place in this scheme. But even with the present version of the theory, some of nature's riddles would be solved. We could explain why the photon has no rest mass energy; and why we do not observe any magnetic monopoles — the magnetic counterparts to electrical elementary charges. These charges, postulated in 1938 by Paul Dirac, could only exist at very large masses. And a final, important possibility is contained in the theory: neutrinos could have a measurable, if tiny, rest mass. And besides this, the sixth basic force and proton decay in the early universe might help us understand why we now look out on an asymmetrical cosmos, filled solely with matter and not with an equal amount of anti-matter.

Chapter IV
Man as a Yardstick
for the Big Bang

Models of the Universe —

Minimal Conditions for Life on Earth

Let us imagine that we could set ourselves outside space and time and that from this vantage point we could regard our universe holistically, consider it in its full extension in space and time, from the Big Bang to the final collapse — or through all eternity, according to our model. Let us assume that we would be able to look at the universe as if it were a sandbox, with each grain of sand representing a galaxy, and take in its overall characteristics. Would we be able to say whether this universe offered good possibilities for the origin of life? Would we be able to specify just what a universe had to look like in order for life to arise, and whether this particular universe matched those requirements?

What in particular would concern us as cosmological omniscients? What would we take into account? As long as we restrict ourselves to the movement of the cosmos on the grand scale, remarkably little data will suffice to give a full description. We would attempt to arrive at values for two figures: (a) the speed at which adjacent grains of sand are flying away from each other (that is, the recessional velocity of neighboring galaxies) and (b) the decrease in this speed over cosmological time — the retardation parameter. In these two magic figures we can more or less crystallize our entire knowledge of the Earth's cosmic environment.

Our present picture of the cosmos dates only from this century, although cosmological ideas can be found in the earliest writings of the Babylonians, Egyptians, Indians and Chinese. Leaving aside the mythological concepts of these early cultures, the first cosmology that might be called scientific was developed by Greek philosophers from the 6th century BC onwards. The modern theories of the cosmos began with a static model of an unchanging universe, suggested by Einstein in 1917. Then, in 1922, Alexander Friedman published in Leningrad the first model of an expanding cosmos. But it was only after Edwin P. Hubble had carried out precise observations of the light from distance galaxies in 1923 that the idea of expansion gained any recognition among scientists. Hubble's measurements showed that galaxies are receding from us faster and faster, the further they are away from us: he had discovered "galactic recession." This finding exactly fitted Friedman's model of the cosmos, and with that, around 1930, the concept of an expanding universe had arrived. Generally speaking, it is still the prevailing theory today.

Nevertheless, it requires one or two subtle steps of reasoning to deduce the expansion of the universe from Hubble's observations. The matter of the cosmos is mainly concentrated in galaxies, islands composed of billions of stars. They typically consist of 100 billion (10^{11}) stars, along with interstellar dust and gas, and take the shape of ellipsoids or flat spirals. These galaxies, in turn, are grouped together in "clusters." The Milky Way — our own galaxy — is one of the "Local Group" of some 20 galaxies, which is probably itself just a section of the Virgo Cluster, comprising around 3,000 galaxies. Galactic clusters represent the dominant form of accretions of matter — at any rate for the visible universe covered by the strongest optical telescopes. Within this part of the universe, astronomy has recorded around 100 billion galaxies, corresponding to an average of one in every sphere two million light years in diameter.

The recession of the galaxies takes on an especially interesting slant if it is followed backwards in time. The further into the cosmic past we go, the closer each galaxy must have been to its neighbor. Consequently, the temperature and density of matter must have been higher as we approach the Big Bang, so that there is no doubt that the cosmic medium was once far hotter and denser than it is now. And a quite inevitable consequence of this is that, in an epoch when the cosmos was very young, radiation and matter formed a very hot plasma — that is, a gas in which all electrons were separated from their atomic nuclei because of the high temperature of the cosmos. Indeed at the beginning of time, it seems there must have been a hot Big Bang — an event in which density, temperature and pressure approached infinity. This is certainly what Einstein's General Theory of Relativity proposes. This explosion, encompassing the entire cosmos, must have happened about twenty billion years ago, about four times the age of the Earth.

Exactly what happened at "zero time," the moment of "creation," is unknown and, as far as this phase of the extreme beginning of the universe is concerned may remain hidden forever; we do not yet know the physical laws prevailing under such extreme conditions. And unfortunately we cannot directly observe the original state of the universe either: not even the notions of space and time are well defined for the beginning of the universe. The electromagnetic remnant radiation we observe — the most ancient historical witness we have yet come across — dates only for 300,000 years after the beginning of the universe.

Each of these two types of expansion establishes a quite definite "geometry" of the universe. An infinite expansion means in geometrical terms that we are living in an "open" cosmos, one that from the very beginning possessed infinitely large cosmic volume — giving a universe with a finite beginning in time but no temporal or spatial end. An expansion destined to reverse, on the other hand, would define a "closed" cosmos, a space with finite volume, which expands from the start to a maximum extent and then shrinks back to zero. Which of these is the actual case is difficult to decide, since the energy of expansion and the cosmic force of gravity are more or less balanced. However, caution is advised in a field where hasty conclusions are always a temptation, especially since many of the "exotic" particles postulated by high energy physics may contribute to the mass of the universe in the form of unseen "dark matter."

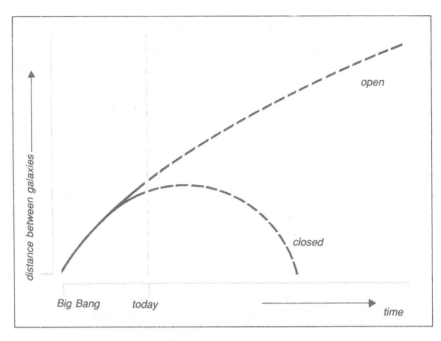

Whether galaxies, stars and life itself can come into being in the universe is decided by the precise value of its speed of expansion. At present it is still uncertain whether this expansion will continue for all time (upper line) or eventually reverse into contraction (lower line).

THE ANTHROPIC COSMOLOGICAL PRINCIPLE

Do these astronomical observations have anything to do with our every day existence on Earth? Are distant and weakly-glowing objects and left-overs of radiation from the Big Bang of any relevance to us? I aim to show that even the remoter regions of the universe have sizeable effects on terrestrial life, and that these are vital to our existence.

Earth and cosmos were linked together in very early cosmologies — by what is today known as a "cosmological principle." In its earliest form, this placed Earth and humanity in the center of nature and the universe.

And it was only under protest that the human race later discarded the idea that we play a central role in the universe. At first, then, cosmologies were geocentric — the Earth was the hub of the universe. In the sixteenth century followers of Copernicus claimed that the hub — at least of the planetary system — was not the Earth but the sun. This viewpoint, too, had to be relinquished once it became clear that our sun is only an insignificant star at the edge of a quite average galaxy, and that even the Milky Way is just one galaxy in a cluster of many thousand.

But in the final analysis, this "unexceptional" position of the Earth is just as much a theoretical assumption that cannot be proved in practice; for the terrestrial astronomer cannot make observations at any appreciable distance from the Earth in order to establish whether the cosmos looks any different from other locations. If he is to convert the observations he has into a model of the cosmos, therefore, he is forced to work on an assumption (see Chapter IX), since his observations, both past and present, derive from the special position of the Earth — whereas, to be able to construct a cosmological model, he needs to know the distribution and properties of matter throughout the entire universe. Cosmology therefore follows the philosophical guideline familiar to other branches of science since the time of the scholastic thinker Occam: select the simplest possible model that will at the same time explain the largest possible number of observations.

Thus, the *Copernican Cosmological Principle*, in its modern version, declares: the Earth does not represent a special observational position. The evaluation of telescope photographs shows that galaxies are more or less equally distributed in all directions. This means that the universe, at least as far as we are concerned, has no favored direction; it is *locally isotropic*. If the Copernican principle is applied to this, the same must apply for all observers — that is, the universe is isotropic as seen from any other point. As a result, all directions are equal and there is no favored point, and in particular no "center." This last characteristic is known as *homogeneity*. A homogenous and isotropic universe has the greatest possible degree of spatial symmetry and the simplest spatial structure.

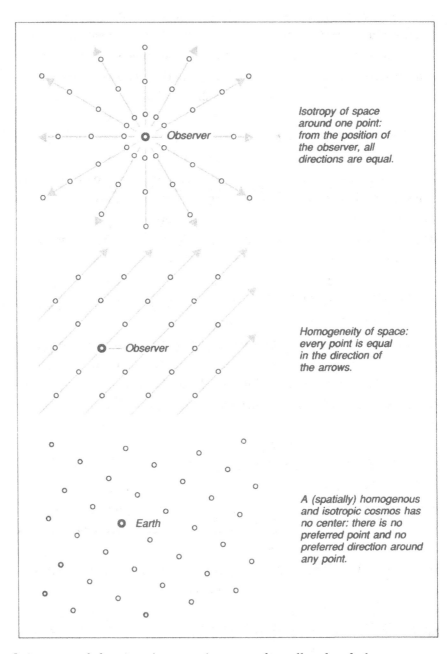

Isotropy of space around one point: from the position of the observer, all directions are equal.

Homogeneity of space: every point is equal in the direction of the arrows.

A (spatially) homogenous and isotropic cosmos has no center: there is no preferred point and no preferred direction around any point.

It is assumed that cosmic space is extremely well-ordered: homogenous and isotropic. The cosmos must have possessed this characteristic right from the earliest moments of its existence.

Isn't this contradicted by the fact that all the other galaxies are flying away from us? This apparent contradiction disappears if we consider the "currant bun model" of the expanding cosmos. Imagine a dough in which currants are distributed with perfect regularity. When the dough is baked, it rises, and the distance from currant to currant increases. If we were to stand on one particular currant, we would have the impression that all the other currants were moving away from us. And we would find exactly the same thing if we chose any other currant — for which we can safely substitute "galaxy." In other words: the expansion has no spatial center.

A number of cosmologists have raised the Copernican principle to the level of a *Perfect Cosmological Principle*, by demanding, in addition to spatial homogeneity and isotropy, a temporal homogeneity — an "invariance" in time. This means that there is no special point in time, no beginning or end to the universe — homogeneity and isotropy throughout all of time. Clearly, there is no room in this version for a Big Bang (among other things), so that models following this enhanced principle (such as the "continuous creation" model) are contradicted by observations that indicate the past occurrence of a Big Bang. One reservation should be added however: in a model postulating an oscillating (pulsating) universe, the Big Bang would appear not as a singularity, but merely as an unexceptional "rebound" event, or "big bounce." Within this model, too, there would be no beginning of time.

A contrasting position is taken by the *Anthropic Cosmological Principle*. This is a restatement of the anthropic principle, as applied to the cosmos. It states that we are observing it at a very special moment in time, and that our present epoch represents a unique chapter in the history of the universe, different from the time before it and the time that is to follow.

In terms of this principle, there is at least one special moment in time — our own. It is marked by the arrival of an intelligent observer. Special conditions had to be satisfied in the time before to permit the evolution of life. And how do the homogeneity and isotropy of space fit into this model? Both continue to apply. If the universe were less ordered and regular, more chaotic than the cosmological principle assumes, there would probably be no life in the universe, and in any event the human race would not have come about.

Proof of isotropy in the immediate neighborhood of the Earth (local isotropy) came in a discovery made by Arno Penzias and Robert Wilson of the Bell Laboratories in 1965, for which they were awarded the Nobel Prize in 1978. Quite unexpectedly, they came across electromagnetic microwaves — radio waves with wavelengths in the centimeter region — that were extraterrestrial in origin. This radiation evidently filled the whole of the cosmos, and was reaching the Earth equally from all directions. Precise measurements taken for every portion of the sky, and more recently confirmed in the infra-red and other wavelengths by observations from planes, balloons, and radio telescopes, showed that this radiation is isotropic to an accuracy of at least one part in ten thousand. It is as if we were "seeing" in this radiation the dying embers of the cosmic fireball — a witness to events of the year 300,000 of the life of the cosmos, when, through its own expansion, matter had become thin enough and cool enough (its temperature having fallen below 4000 degrees) to become transparent to its own heat radiation. This cosmic background radiation is the electromagnetic echo of a late stage of the Big Bang, with a present-day temperature of 3 degrees Kelvin — 3 degrees above absolute zero. Although this had been qualitatively predicted in 1948 by the US physicists George Gamow, Ralph A. Alpher and Robert Herman, neither Penzias nor Wilson were aware of the prediction in 1965.

Why It Gets Dark at Night

Children, and many adults, no doubt suppose that darkness falls on the Earth at night because the sun is then shining on the other side of the world and our side is in shadow. This could hardly be more wrong. The correct explanation is quite different. It gets dark at night because we are living in an expanding universe — a universe that began with a Big Bang.

George F. R. Ellis of the University of Cape Town put his finger on the problem when he remarked: "While photographs of astronomical objects reveal many striking phenomena, the single most important feature of these photographs is the fact that *the sky is dark* between the images of galaxies, quasi-stellar objects, and other dramatic objects."

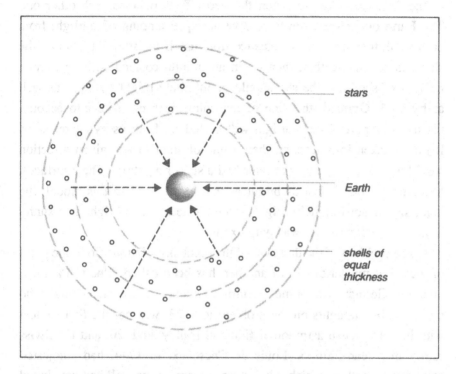

Olbers' Paradox in an infinite universe. Equal amounts of light reach Earth from each shell of stars; in an extreme case, the night sky ought to be as bright as day. It was vital to the existence of life that the sky was dark at night.

That this poses a problem was already clear to Johannes Kepler. In 1610 he wrote, in his book ".... If they [the stars] are suns having the same nature as our own, why do these suns collectively not out distance our sun in brilliance?" He, too, marvelled in his time that it was not equally bright during the night as during the day. A moment's thought seems to point to the same conclusion. Taking the simplest model of the universe, we find the Earth surrounded equally on all sides by stars. If we imagine these stars to be arranged in spherical shells of equal thickness spreading outwards from the Earth, then the number of stars in each shell which is directly proportional to the surface area of each shell will increase as a function of the square of that shell's distance from Earth. At the same time, however, the intensity of starlight falls with the square of the distance; taken together, the two effects balance each other out. The Earth ought therefore to receive an equal amount of starlight from each shell, regardless of its distance from us! If the starlight from all the shells is added together, then — in an infinite cosmos with an infinite number of shells — the sky should be infinitely bright by night as well as by day! Granted, this line of reasoning does not take into account the fact that part of the starlight will be hidden from us by nearer stars. If this is taken into account, then a line of sight in any given direction would nevertheless sooner or later find a star or a galaxy. Their surfaces, however, would be just as bright as that of our sun; in this model, the night sky, if not infinitely bright, should at least be as bright as a sunny day. No wonder Kepler was perplexed!

The problem of the darkness of the night sky, obvious from one point of view but mysterious from another, has been called Olbers' Paradox, after the German astronomer Heinrich Olbers. However, Olbers, who presented his thoughts on the problem in 1823, was not the first to deal with it. The British astronomer Edmund Halley, in 1720, and the Swiss astronomer Jean-Philippe Louis de Cheseaux, in 1744, had suggested solutions — both of which, like Olbers', were wrong. All had concluded that the night sky was dark because interstellar gas and dust absorbed the rays of light from distant stars. But this proposal — as John Herschel noted in 1848 — cannot explain the phenomenon, for the simple reason that the absorbing gas would itself heat up and eventually emit just as much radiation as it absorbed.

In an expanding cosmos with a finite past, the "paradox" disappears. Firstly, the light from distant stars and galaxies, which are receding from us, reaches us shifted towards the red end of the spectrum and therefore weaker in energy. Furthermore, the universe had a finite beginning, and stars have only a finite life-time; thus, a given distance away (= a given time in the past), there are simply no more galaxies to put out light. The question was finally put to rest when in 1987 Cambridge astronomer Paul S. Wesson together with Knut Valle and Rolf Stabell of Oslo University had worked out the relative importance of the two factors. They concluded: "Expansion of the universe only dims the light by a factor of about 2." And stated "once and for all" that the night sky is dark "because the universe is still young, not because it is expanding."

IF IT DIDN'T GET DARK...

The older the universe is and the slower it is to expand, the brighter the night sky becomes. Not only the brightness of starlight, but also the temperature of the cosmic background radiation — at present 3 degrees Kelvin — would become higher. In the recontraction phase of a closed universe, the intensity of starlight and the background temperature would constantly increase; and in the late phase of a cosmic collapse these would eventually wipe out all life on Earth.

This brings us to an important question. How hot and how bright is it permissible for the night sky, and with it the cosmic radiation background to have been four billion years ago if they were not to arrest the development of life on Earth? We might suppose that a brighter night sky could lose us some sleep, but we would be able to live with it. But it is not that simple — as we can see if we consider the full implications.

From a thermodynamical point of view, life on Earth exists on a "slope"; (free) energy flows down this "slope" towards us, is consumed and thereby converted to useless heat, which is given off again. At the top of this hill sits the sun, supplying energy, and at the foot of the hill we can imagine the darkness of space, receiving the waste heat. Life presents a metastable state imbedded in a thermodynamic non-equilibrium. The Earth gets its energy from the sun, whose surface radiates at a temperature of 6,000 degrees. This energy drives all processes on Earth: weather, climate, plant photosynthesis, etc. After use, the remaining energy is radiated back into space as heat waste in the form of infra-red light at room temperature (20° to 30°C). High-grade energy reaches us from the sun and is given back into space as low-grade energy.

THE COSMIC CALENDAR

COSMIC TIME	ERA	TIME BEFORE PRESENT
Zero	Singularity	20 billion years
10^{-43} sec.	Planck Time	20 billion years
10^{-6} sec.	Era of Hadrons	20 billion years
1 second	Era of Leptons	20 billion years
1 minute	Era of Radiation	20 billion years
1 week	Era of Radiation	20 billion years
10,000 years	Era of Matter	20 billion years
300,000 years	Era of Decoupling	19.7 billion years
1-2 billion years		18-19 billion years
3 billion years		17 billion years
4 billion years		16 billion years
4.1 billion years		15.9 billion years
5 billion years		15 billion years
10 billion years		10 billion years
15.2 billion years		4.8 billion years
15.4 billion years		4.6 billion years
15.7 billion years		4.3 billion years
16.1 billion years	Archaeozoic	3.9 billion years
17 billion years		3 billion years
18 billion years	Protozoic	2 billion years
19 billion years	Palaeozoic	1 billion years
19.4 billion years	Cambrian	600 million years
19.55 billion years	Silurian	450 million years
19.6 billion years	Devonian	400 million years
19.7 billion years	Carboniferous	300 million years
19.75 billion years	Permian	250 million years
19.8 billion years	Triassic	200 million years
19.85 billion years	Jurassic	150 million years
19.9 billion years	Calciferous	100 million years
19.95 billion years	Tertiary	50 million years
19.97 billion years	Tertiary	30 million years
20 billion years	Quartenary	1 million years
20 billion years		120,000 years
20 billion years		50,000 years
20 billion years		2,000 years

The "Cosmic Calendar" maps out the most important events in the development of the universe over the timescale of a single year. The Big Bang is thus placed

EVENTS	COSMIC CALENDAR (20 bn yrs = 1 yr)
Big Bang	January 1, 0:00
Formation of particles	January 1, 0:00
Proton-Antiproton annihilation	January 1, 0:00
Electron-Positron annihilation	January 1, 0:00
Formation of deuterium and helium	January 1, 0:00
Radiation thermalized	January 1, 0:00
Matter dominates	Jan. 1, 0:00 + 15 sec.
Universe becomes transparent	January 1, 0:08
First galaxies	January/February
First galaxy clusters	February
Our proto-galaxy collapses	March
First stars	March
Oldest quasars, stars of population II	April
Stars of population I	June
Formation of proto-solar nebula, Sun	October
Origin of planets, first rock	October
Lunar craters formed	October
Oldest rocks on Earth	October
First microbes	November
Oxygen in Earth's atmosphere	November 25
Microscopic life forms	December 12
Algae, sea life	December 19
ldest land plants, first vertebrates	December 22
Fish, insects, early ferns	December 23
Forests, coal deposits; amphibians	December 25
Reptiles	December 26
Dinosaurs; continental drift/first mammals	December 27
Birds	December 28
Alpine folding	December 29
Vulcanism	December 30
Primates	December 31
Homo sapiens, mammoths	December 31, 11:15 pm
Neanderthal Man	December 31, 11:55 pm
Cave paintings	December 31, 11:58 pm
Birth of Christ	December 31, 11: 59:57 pm

at 0:00 on January 1 and the present day corresponds to midnight on New Year's Eve. On this scale, 1 month represents 1.7 billion years, 1 day = about 50 million years, a half hour = 1 million years, 1 minute = about 40,000 years and 1 second = about 700 years

The steep temperature gradient between the sun and the surrounding sky is a precondition and continuing mainspring for the functioning of the terrestrial biosphere. To be precise, it is the temperature difference between the sun and the rest of the sky, so far as this is not obscured by the sun itself: that is, both the day and the night sky serve as waste pits for our unwanted heat. This heat flow defines the non-equilibrium that keeps our lives in operation. If the night sky were to glow with the brightness of the sun, the surface of the Earth would heat up in the course of time to about 6,000 degrees — and then, instead of a temperature gradient, a thermal equilibrium state would prevail on the planet; the same amount of heat would be radiated into space as was received from the sun and the rest of the sky. Under these conditions, life could never arise. And the situation would have a further catastrophic consequence, this time for the stars themselves: in a cosmic "heat bath" as hot as the surface of the stars, even the temperature of the stars would rise and after a time they would vaporize or explode. They would dissipate and be lost into the hot cosmic gas.

So the temperature difference between the sun and the rest of the universe must not fall below a certain value. "The Earth would be a death planet if the temperature of the night sky was only 300°K [80°F]," writes Ellis. While stars would not vaporize at this temperature, the exchange of "free energy" into heat on the Earth would still be too little. Thus, the expanding universe with its finite past permits a local non-equilibrium state to be maintained on the Earth. In short, it also follows from the anthropic principle that it must get dark at night!

This condition would also set limits to the life-time of a closed universe. Here, the temperature of the cosmic background would reach its lowest value at the moment of maximum expansion. The shorter the period of expansion, the higher this minimum would be. And in order for the minimum to be only a few degrees Kelvin, close to absolute zero, a closed cosmos must have a total life-time of at least thirty billion years.

LEFT-OVERS FROM THE BIG BANG

Through its far-reaching forces and its ubiquitous radiation, the cosmos exercises a direct and lasting influence on events on Earth, permitting and to some extent controlling the development of life on its surface. This also goes for cosmic microphysics, that is, the chemical elements formed in the Big Bang. The journey backwards in time to the earliest stages in the history of the universe is not unlike a journey through the most recent history of the physics of elementary particles. The closer we approach the beginning of the universe, the hotter (richer in energy) and denser the cosmos becomes. Particle physics too has progressed to higher energies in recent years, discovering new particles in the process. Some of the events simulated in the large accelerators have not occurred since the first moments of the Big Bang.

The closer our theories and experiments bring us to the very beginning, the smaller the region that a hypothetical observer could observe becomes. The part of the universe observable at any time from a given point comprises a region as large as that which could have been crossed by light in the time since the Big Bang. The limits of our description of the Big Bang are drawn at the point where Heisenberg's uncertainty principle and quantum mechanics make the concepts of space and time meaningless — where even Einstein's theory of gravitation is no longer valid as a theory. This limit is defined in time by what is called the Planck time, 10^{-43} seconds after the Big Bang. The density of matter at this point in time is still fantastically high: there would be 10^{94} grams in every cubic centimeter (10^{84}) tons / cubic foot, compressed at the unimaginable temperature of 10^{32} degrees.

These are the conditions prevailing at the earliest moment after the Big Bang that we can still make "sensible" statements about: as far as our scientific understanding goes, then, it is "the beginning of time." What physicists envisage in their grand unification theories was then reality: all of the forces of nature were equal in influence and strength. In this almost infinitely hot radiation bath, particles were constantly coming into being and disappearing again. However, rapid expansion brought with it rapid cooling, so that more and more particles — first the heavier ones, then the lighter — "froze out," that is, they were created but not annihilated; they became stable, and no longer destructible by high-energy photons (gamma rays).

The first millisecond also saw the creation of *gravitational waves,* or gravitons. In the same way that there is a cosmic background radiation composed of photons, it is expected that there is a cosmic background radiation of gravitational waves: according to calculations in the models so far proposed, they should have a temperature of around 1 degree Kelvin and a wavelength of about 1 millimeter. Direct evidence of gravitational waves, to be achieved using detectors currently at a development stage, including Doppler-tracking of interplanetary spacecraft, is not anticipated before the end of the century. From the point of view of the anthropic principle, this "weakness" of gravitational waves is also their "strength" — for even if all the other conditions necessary for life were satisfied, our "island Earth" would also need to be left as much as possible in peace by the unceasing interstellar to-and-fro of chaotic gravitational forces. The best "development aid" cosmic gravitation has given to terrestrial life is our constant orbit around the sun.

The strength of various kinds of radiation striking the Earth must be limited; this applies to cosmic rays, X-rays and gamma rays as well as to gravitational waves left over from the Big Bang. If these were much stronger, they would make life on Earth decidedly uncomfortable, or even wipe it out altogether: it would be like living in a zone of continuous Earthquakes. No one would be safe living in or near a building — if it were possible to build any in the first place. And it would be impossible to practice science in the way we do now. Heavy items would turn up in different places from day to day, and their movements could not be predicted. Any laboratory would be forever changing, making experimentation a hopeless activity. Gravitational waves might even be so strong they would break the Earth apart. The fact that we can reliably predict the motion of the Earth with the natural laws we have produced, however, shows that the cosmic background of gravitational waves from the Big Bang is no bother to us — a consequence of gravitational forces and the special physical conditions in the early universe.

In the first millisecond of the universe there must also have seethed a hot soup of quarks and anti-quarks: this, at least, is what quark theory predicts (see Chapter III). At temperatures around 10,000 billion (10^{13})

degrees Kelvin these quarks would be created in pairs and immediately wiped out again. At lower temperatures, cooled by cosmic expansion, no new quark pairs would have come into being, and those remaining would have met in mutual annihilation. All that remains is a small remnant of *primordial* quarks, so sparsely distributed that they no longer destroy each other. If these could be spotted and observed, it would be a further success for the theories of the Big Bang and of elementary particles; and a researcher by the name of Fairbanks in Stanford, California has been endeavoring to do this for some years. Early in 1981 he announced he had found traces of particles with a fraction of the elementary charge. Primordial quarks differ from those indirectly observed in accelerator experiments in that they are "free," rather than being seemingly inseparably welded together inside a neutron or a proton.

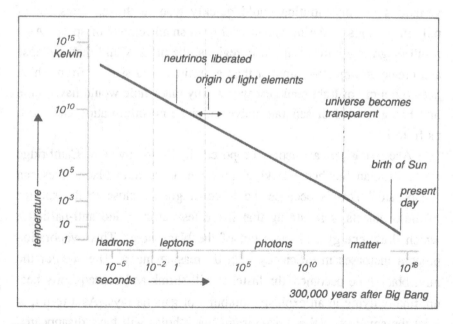

The development of the cosmos from the Big Bang to the present day, as charted by its temperature curve.

As we saw in Chapter III, the nuclear force between quarks is like a rubber band, slack when the quarks are close together, taut when they move apart. However, as the nuclear forces are effective only over the very short distance of 10^{-13} centimeters, it is quite conceivable that at greater distances the rubber band might snap, so to speak, and the quarks might then appear independently and singly. On the basis of this theory it has been estimated that there ought to be one free quark to every billion hydrogen atoms — which is about the same frequency of occurrence as that of gold among all the elements of the cosmos. Studies of sea water, however, have given an upper limit of at best one quark to every 10^{20} hydrogen atoms — which could give rise to doubts concerning the current model of the Big Bang. Was it ever as hot as 10^{13} degrees in the Big Bang? Where the fault lies will have to be clarified by future research.

A third possible left-over from the first millisecond would be mini black holes. In conditions of hot and dense matter and radiation, minor variations in concentration could quickly lead to the compression of millions of tons of the mass and energy in small regions of space. As a result of gravity built up in these regions, the mass would then collapse and create an object — smaller than a millimeter in size — from which not even a ray of light could escape. A tiny black hole would have come into being, a kind of separate universe, since no information can reach us from it.

According to a theory proposed in 1974 by the Cambridge mathematician Stephen Hawking, however, these mini black holes can "evaporate." This is because the force of gravity close to the surface of these objects is so strong that it creates particles and anti-particles, which draw energy and mass out of the black hole. This evaporation process increases in efficiency with decreasing mass. The smaller the mini black hole becomes, the faster it will shrink and consequently heat up, finally dying in an explosive outburst of gamma rays. As a result, at least the smallest of these *primordial* black holes will have disappeared from the cosmic arena shortly after their birth. A hole with the mass of one million tons would have evaporated after only thirty years, while a hole of one billion tons would last for just three hundred million years. Most of the mini black holes originating in the Big Bang, then, would have vanished long before now; only those with more than 4 billion tons of mass — the mass of a small mountain — would have lasted till the present. Astronomers were watching for their final flare-ups but found nothing. In addition, the cosmic background radiation in the gamma rays sets a strong upper limit to the possible abundance of such objects.

How many of these postulated mini black holes might have been produced in the Big Bang depends on how irregular, how chaotic the universe was; the more irregular the density, the easier it would have been for matter to collapse in particular regions. As we shall see, however, the anthropic principle allows us to demand that the universe must have been very symmetrical — homogenous and isotropic — very close to the Big Bang, otherwise there would probably have been (among other things) no galaxies in which stars could later form.

HYDROGEN AND HELIUM: THE RAW MATERIAL OF STARS

The Big Bang was responsible for the production of the raw components for the chemical elements from which, much later, life could develop. As in a good recipe, however, what mattered most was the exact amount of each ingredient in the overall mixture. In its very first minutes, the Big Bang only produced the lightest elements of all: hydrogen (with one proton in its nucleus), the heavier hydrogen isotope deuterium and finally helium. Deuterium — with one proton and one neutron in the nucleus of its atom — is very rare: on average only one deuterium atom is encountered for every 10,000 hydrogen atoms. Unlike other elements, the abundance of deuterium we observe must have existed since the Big Bang itself. In the fiery furnaces of nuclear fusion existing in the interiors of stars, this fragile element would have been destroyed at once. For this reason, deuterium is a fine pointer to the nature of some of the processes of the Big Bang and their effects on our existence today. Deuterium also represents a transitional stage in the nuclear fusion of hydrogen to helium. In the Big Bang about a quarter of all hydrogen was transformed into helium. How did this come about?

101

Let us go back once more to the end of the first cosmic millisecond. The era of nuclear forces, the "hadron era," had just finished, and the heaviest elementary particles had destroyed each other in pairs apart from a small remainder so thinned out by the expansion that they were no longer able to come in contact with one another. Besides these there were neutrinos, reacting via the weak nuclear force with the protons, electrons and positrons. In this phase, which is also known as the "lepton era," specific quantities of neutrons and protons also came into being. The temperature at the end of the first second was 10 billion degrees; there were free protons and neutrons swirling in the cosmic gas. Time was short, since free neutrons decay after 15 minutes. At this point, however, it was still too hot, and the neutrons were carrying far too much energy for stable atomic nuclei to be able to form. This took place after about one minute, when the first cosmic nuclear fusions began. By then, the temperature had decreased by a factor of 10 to one billion degrees. Protons captured neutrons, deuterium was formed, and the capture of a further neutron by each nucleus resulted in the formation of the unstable hydrogen isotope tritium. The tritium nuclei reacted with free protons and converted to helium. In this way, nearly all the available free neutrons were used in the making of helium nuclei; and the cosmic gas contained not only hydrogen but also 27 percent helium — helium that had been produced in the Big Bang. By this point, the universe had already cooled so much that nuclear fusion could no longer take place to produce heavier elements. It was left to the stars to carry this out.

What roles does the anthropic principle play in this scenario of the origin of the elements in the cosmos? The number of helium atoms produced at the end of the reaction chain was determined first and foremost by the temperature of the gas. Prior to this, the temperature was also the deciding factor in the number of protons and neutrons that "froze out" at the end of the lepton era — a process that occurred once the cosmic temperature fell below that typical to the reaction. And the speed of cooling was directly determined by the speed of cosmic expansion.

The two processes were in competition with each other — cosmic expansion cooled the universe down, thus fixing the length of time in which a particular temperature-dependent particle reaction was possible. The speed of expansion is regulated by gravity, so it is gravity that indirectly determines the length of this reaction period: the weaker gravitation is, the quicker the expansion will be and the shorter the reaction period. How many reactions can fit into this period, on the other hand, depends on the strength of the weak nuclear force: the stronger it is, the more protons react with electrons and neutrinos and convert them into neutrons. And finally, the number of reactions possible also depends on the particle density at any point in space.

The reaction speed and the cooling rate, then, had to assume a quite specific relationship in order to produce just the abundance of hydrogen, deuterium, and helium we observe. The weak nuclear force and gravity had to act against each other in such a way that the incidence of helium did not exceed 30 percent: the relationship, in fact, had to be $(alpha-W)^4 = alpha-G$.

Consider the effect of variations on the relationship. If all protons had been used up in the production of helium, the eventual result would have been a total absence of water (which is, after all, nothing but oxidized hydrogen). Without water as a chemical solvent life — at least, life as we know it — would never have been able to come about. That is what would have happened if gravitation had been weaker relative to the weak nuclear force: the universe would then have expanded too quickly for neutrons to convert into protons by weak interactions. As a consequence, the original 50:50 mix of protons and neutrons would largely have been retained and nearly all protons would have found partners so as to fuse into helium. As a further consequence, stars could have formed exclusively from helium gas; these are known to have a significantly shorter life-time than hydrogen-burning stars like our sun — probably too short for the evolution of intelligent beings which requires at least a few billion years, even if a comparable solvent could stand in for the missing water (see Chapter VIII).

Prospects are no better in the opposite case. If gravity had been stronger the universe would have expanded too slowly: most of the neutrons would then have been converted into protons by interacting via the weak nuclear force. As a result there would have been fewer neutrons and hardly any helium would have been created. Whether this situation would have prohibited the development of life is not as clear an outcome as in the first alternative. Until we know more in this area, the anthropic principle is unable to supply any quick answer here.

THE MASS OF THE NEUTRINO

The cosmic incidences of hydrogen, deuterium and helium clearly constitute a sensitive "adjusting screw" that can only be "turned" a very few degrees without putting a question mark against biological evolution. Similar limitations apply to the strengths of the fundamental constants — in their absolute values and relative to one another — that were involved in the processes by which these elements came into being. And similar conditions hold for neutrinos, those particles of the lepton family whose mass was considered to be "probably zero" as late as 1980. A neutrino mass differing from zero would have appreciable cosmological and anthropic consequences, as it has an immediate effect on the total mass of the universe. This is because a vast number of neutrinos are understood to have been born in the Big Bang — about 250 million to every baryon. This gives a total of about 10^{88} neutrinos — a quarter of the number of photons in the cosmos. With this great a number of neutrinos, even a small neutrino mass would considerably increase the mass of the cosmos. It would also mean that matter was denser during the lepton era than hitherto assumed, and there would then be more neutrons and as a result more helium. And finally, a non-zero mass for the neutrino could make the mass of the universe so great that it would be bound eventually to re-collapse, making for a closed universe rather than an open one. This, in turn, raises the question of the "missing matter." The problem of the "missing matter" is not a new one, and arises because astronomers have found less visible matter in the universe than there "really" should be. The reasoning behind this is that most forms of matter make themselves apparent through one or another type of radiation, from radio waves through infra-red, visible, and ultra-violet light to X-rays. As we have a reasonable idea which objects are emitting this radiation (in the main, stars), the strength of the radiation can be converted by calculation into quantities of matter.

This mass, however, can also be calculated in another way — at least for "gravitationally bound" systems (systems held together by gravity): binary stars and galaxies or clusters of galaxies. And here a problem arises, particularly in the case of galactic clusters, which generally comprise several thousand individual galaxies, all in random motion relative to one another. This independent motion of the member galaxies would have dispersed the cluster long before now if it were not held together by the gravity acting between each of them. But a comparison of the two methods of establishing the mass content showed that at best 10 or 20 percent of the gravitationally bound matter is also optically visible as stellar matter in clusters of galaxies. For the amount of matter alone to arrest the expansion of the universe (and thus create a re-collapsing one instead of an ever-expanding one), the amount of matter observed, according to current calculations, falls short by a factor of 10 to 50.

The dilemma may be formulated either as a problem of "missing matter" or as one of "missing radiation." Speculations are rife regarding the mysteriously invisible remainder. Cold intergalactic gas clouds have been proposed, as have black holes — the remains of stars that have collapsed under their own weight. These objects can in principle contain a good deal of mass without its necessarily becoming noticeable through optic radiation. Light was shed on this dark corner of cosmology in 1980 from an unexpected direction: the physics of neutrinos.

The elementary particles known as the neutrino — "the little neutral one" — fully deserves its name: a particle with no charge, no magnetic moment and — as assumed until recently — no mass. It was first observed experimentally in 1956 at the nuclear reactor in Savannah River, Georgia.

Since then the neutrino has, so to speak, assumed a split personality and put on weight. On the basis of the theory of quarks and leptons (see Chapter III), it is now believed that there are three types of neutrinos — the electron neutrino, the muon neutrino and the tau neutrino.

Experiments carried out have indicated that the neutrino may possess a small mass after all, and that it may be able to "oscillate" from type to type.

At the end of April 1980, three physicists at the University of California at Irvine, Fred Reines, Henry W. Sobel, and Elaine Pasierb, announced their finding that electron (anti-) neutrinos had changed type en route from the heart of the Savannah reactor to a detector 11.2 meters away. Their experiment, they claimed, indicated a mass for the neutrino of at least one electron volt (equivalent to 2×10^{-33} gram). A Moscow research team confirmed the American claim very shorly afterwards: according to them, the neutrino mass was between 14 and 46 electron volts. This is about one hundred-thousandth of the mass of an electron.

Observations at CERN in Geneva (Switzerland) in 1979 had already established that muon neutrinos did not convert to electron or tau neutrinos; if this sort of event happened at all, it had to be a case of electron neutrinos oscillating to tau neutrinos.

One question immediately arose at this point: why has the mass of a particle that has been familiar to scientists for a good twenty-four years still not been measured?

The reason it is so difficult to measure the mass of the neutrino lies in the fact that, in radioactive decay or events in a nuclear reactor, the other energies involved in the reaction are so much greater than the tiny rest mass of the neutrino that it can not be separated from background static noise. But the idea that the neutrino might have mass is by no means new. It was first discussed by a group of Japanese physicists in Nagoya as long ago as 1963, and neutrino oscillations were first considered by Bruno Pontecorvo (Dubna, USSR) in 1968. All these questions were rejected at the time, because there was then no experimental evidence for a neutrino mass.

All the more remarkable, then, are the possible consequences of a neutrino mass for the problem of the missing matter. In an exemplary case of bridge-building between disciplines — particle physics and astrophysics — David N. Schramm of the University of Chicago and Gary Steigmann of the University of Delaware were able to estimate what the neutrino's mass would be.

The basic idea was simple. Given their large number, the 10^{88} primordial neutrinos could significantly increase the mass of the cosmos over that so far observed even if the mass of each neutrino were tiny.

Since primordial neutrinos, interacting only through the weak force, are not open to direct observation, Schramm and Steigmann suggested they should be investigated through their indirect effects — namely, their interaction with gravity. Like anything else, neutrinos, if they possess a mass, will be attracted by stars and galaxies, and should thus become available to observation. The investigator's conclusion was that neutrinos with a mass between 3 and 10 electron volts (about 10^{-32} grams) would be in a position to contribute the lion's share of the mass of the universe. Neutrinos, they decided, would mostly be drawn into galactic clusters by the force of gravity.

What becomes clear from a careful examination of the problem of the missing mass is that the discrepancy between visible and gravitationally bound mass only arises in a serious way in the case of galactic clusters. Mass is also "missing" in smaller systems; but there the problem can be solved by assuming the presence of black holes and the like. In galactic clusters, however, the invisible part cannot be hidden in nucleons: this would contradict the implications of the observed abundances of cosmic deuterium and helium. Here the primordial neutrinos offer us a natural way out.

The problem of the missing mass does not occur in the case of galaxies, binary stars or star clusters. Here, it is true, we only see a part of the mass; but this can be explained by means of the assumption that the invisible remainder is, in the form of nucleons, cold gas clouds or black holes. Since only neutrinos heavier than 10 electron volts would be heavy enough to be captured by individual galaxies or other smaller systems the neutrino mass should be *less* than 10 electron volts.

But there is also a lower limit for the mass: in order for the galactic clusters, at least, to be able to retain them, the captured neutrinos must weigh at least 3 electron volts. Taken together, this would give us a "permissible" range of between 3 and 10 electron volts.

The implications of this are considerable:

107

— The problem of the invisible mass in galactic clusters could be solved by supposing neutrinos with mass. The 10^{88} neutrinos would account for about 10^{24} solar masses, almost as much as the known visible matter of the cosmos.

— The question of why Raymond Davis of the Brookhaven National Laboratory, who is attempting to capture neutrinos from the interior of the sun, has "seen" fewer neutrinos by a factor of 3 than predicted, could be answered by invoking neutrino oscillations. On their way from the Sun to the Earth, two-thirds of the electron neutrinos would have changed into tau neutrinos.

This could be tested by improving the reactor results achieved so far; and also perhaps by observing neutrino pulses emitted in future supernova outbursts. On account of their mass, neutrinos would propagate at less than the speed of light; measurements of the time they take to reach us would then determine their mass.

The consequences of neutrinos with mass for the origin of life lie in the influence they may have on the birth of galaxies. A "bath" of neutrino particles filling the cosmos could hardly avoid being a considerable obstacle to the formation of young galaxies. At present, however, there is little precise knowledge of this area.

ORDER VS. CHAOS: WHY IS THE UNIVERSE ISOTROPIC?

To recapitulate briefly: two things at least must coincide in order that an observer may exist in a given universe. First, the laws of nature must allow for the *possibility* of life; and to this structural potential there must then be added the actual opportunity for this possibility to be realized in the evolutionary process. A universe that is to become a self-cognizant universe must acquire the necessary conditions, the material ingredients and the appropriate timescale: in our own universe, this has clearly been achieved on the surface of the third planet in our solar system. How often this same development may have come about elsewhere (perhaps at an earlier or later time) depends on too many unknowns for us to be able to predict even statistically with any degree of credibility. But a highly developed life form cannot have arisen very often: we have yet to come across any signs of a technological civilization — such as radio signals, artificial infra-red stars, or biological messages — despite twenty years of searching. This makes it likely that mankind is the sole observing intelligence, at least in the immediate cosmic neighborhood. Had the birth of the planets taken, even in minor details, a different course our own universe too might have had to get by without us. Looking ahead to the distant future of the cosmos, it is still probable that this will one day be the case.

The problem of the *special conditions* and the courses of evolution leading to the origin of life brings us back to the Big Bang and the initial conditions particular to it, from which all else developed. The beginning of the universe is not open to direct examination, and the best witness remains cosmic background radiation. A few minutes after the Big Bang, at the close of the era in which the elements originated, the cosmos entered a phase with less incident in which electromagnetic radiation was mixed with matter in a sort of cosmic primeval soup, which is often called the "cosmic fireball" — a misleading name, as there was never a ball isolated in space to be filled with the hot gas of the early universe: the entire expanding cosmos was always evenly filled with it.

The radiation era during which radiation and gas were in equilibrium, ended after about 300,000 years. After this the fog lifted: protons and electrons united to form neutral hydrogen atoms, and the gas became thin and cool enough to be transparent to electromagnetic waves; from then on, they penetrated practically without hindrance throughout the expanding universe. This radiation thus reaches us today virtually unchanged. From its last "contact" with the cosmic gas, it carries with it information about the spatial distribution of those days. Any variation in density in the year 300,000 PUN (Post Universum Natum = after the birth of the universe) would be observable today as a variation in the background radiation dependent on the observation angle. However, within the accuracy of the observing instruments (99.9%) no such variation has been observed.

The consequences of this finding are far-reaching and influence our model of the cosmos; and it also throws light on a strange fact underlying our existence — even though there is no apparent reason why we should be concerned about whether the cosmic background radiation reaches us evenly or irregularly: after all, its weak radio waves do not even disturb our broadcasting!

What is so singular about isotropy and homogeneity is that they imply that the early cosmos must have been very "symmetrical" and regular.

Among the various hypothetical possibilities as to how the universe could have started off, this is probably the most unlikely beginning! A more likely and plausible one, and one that scientists would therefore find easier to swallow, would be a beginning that was relatively unpredictable, highly irregular and in chaotic disorder. This would be "reasonable" and hardly need any further explanation. From this mythical "primeval chaos" the universe could assume a present state that could be completely independent of initial condition. This would be like a ball thrown into a bowl: it will always come to rest at the lowest point, in the center, regardless of where it first lands. In the words of Britain's W. H. McCrea: "What we observe now does not depend on the way it started." The relativity theorist Charles W. Misner (University of Maryland) had just such a primeval chaos in mind when he put forward his chaotic cosmology in 1967. He presented a mathematical model of the primeval chaos to which he gave the name "Mixmaster Universe": just as a food mixer scrambles the ingredients into a homogenous soup, the initial disorder was to be smoothed out in his model in the course of cosmic expansion.

The critical test of such a model is that it must show how the primeval disorder could have been smoothed out, "damped down," by the year 300,000 PUN. Otherwise it would be contradicted by observations. The ironing out of the cosmos might have been effected by a number of different mechanisms: neutrinos, additional particles created by the irregular gravitational field, mini black holes. These might have drawn energy from the chaos and so evened out the irregularities; the universe would then have been left in an ordered, isotropic and homogenous state.

But this idea only worked out to a limited extent: the chosen method of smoothing out the chaos, like friction, would have produced considerable heat radiation and entropy, and the cosmic background radiation would then have had too strong an entropy and look different in the microwave range. A cosmic primeval chaos, therefore, could only be partly ironed away by these means.

What seems to have been the final rejection of the idea of a primeval chaos was formulated in 1973, when the Cambridge physicists C. Barry Collins and Stephen Hawking showed that any universe with an uneven initial condition would be bound, after initially evening out, to grow uneven again as time went on. This would happen no matter what "friction" mechanisms tended to smooth out the chaos — and these would in any case have lost their effect after the first fraction of a second. The inevitable conclusion from this is that the universe must have developed from an extremely even initial condition as early as 10^{-35} second PUN.

How do we explain these most improbable initial conditions? The danger here is that we may easily be led to misuse unnecessarily improbable assumptions about the initial state of the cosmos simply to hide gaps in our knowledge. It is not for nothing that physics has a principle after the style of Occam's razor — "Make use of as few hypotheses as possible" — which prescribes that the fewest possible initial conditions should be assumed. "When postulating initial conditions we must always beware of falling into the trap that ensnared Bishop Ussher in the 17th century. Ussher declared that the universe was created complete in every detail in the year 4004 B.C. Those who later pointed out that fossils exist which are much older were told on good authority that the fossils were also created with the universe." (Harrison) This anecdote, which can now be regarded as no more than a curio, can still serve us as a warning against "resorting to initial conditions as an easy substitute for explanations" (Harrison).

There is another odd angle to the very special initial conditions in the cosmos: the universe "needed" irregularities at a later stage, otherwise galaxies could not have developed. Galaxies owe their origin mainly to intensified irregularities in the distribution of matter.

The first galaxies came into being after one to two billion years. The irregularities from which they formed, therefore, must have arisen at around 300,000 PUN and constantly increased from that time on. The sensitive adjust-ing screw that selects regularity or irregularity for the cosmos is once again the velocity of expansion. It can only be varied between very tight limits:

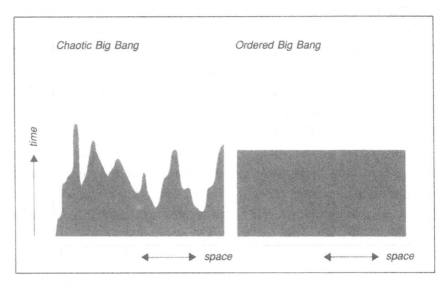

The Big Bang: chaos or order? Astronomical observations of the cosmic background radiation indicate that in the early life of the universe space was highly ordered — isotropic and homogenous (see right). This fact, indicating a special and unlikely initial state for the universe, is still not understood: why should different areas of space behave in the same way very early, expanding at the same rate for example, although there is absolutely no causal connection between them? A more likely condition would be an initial chaos (see left), which would be quickly smoothed out; however, the hypothesis of an initial chaos has not as yet enjoyed much success. A special chaotic model of the universe, Charles W. Misner's "Mixmaster Universe," is contradicted by observations.

— Too fast an expansion of the universe would prevent galaxies from ever forming at all. Irregularities such as accretions of matter would be held back from any effective growth: the expansion would drive the matter apart, just as a strong wind will disperse a mist before it forms clouds.

— On the other hand, too slow an expansion would prevent the universe from growing even to its present size: it would soon fall back upon itself. The difference as far as the early phase is concerned is very small: if the velocity of expansion at the end of the first second were just one trillionth (10^{-12}) slower, the universe would have re-collapsed by the end of only fifty million years. In so short-lived a closed cosmos the temperature would never have fallen below 10,000 degrees.

These exact parameters for the velocity of expansion are the most precise requirement — a sort of fine-tuning of constants — of those we have been able to arrive at by applying the anthropic principle.

We are thus confronted with the fact that the universe obviously expanded quickly enough not to have re-collapsed after only a short time. Our existence demands a supply of gravitationally bound systems — galaxies and stars — and sufficient time to allow biological evolution to take place. In a universe that was collapsing at an excessive rate, the time would be too short; whereas in a fast-expanding universe even regions with a relatively high concentration of matter would be unable to combine into proto-galactic nebulae. Hawking therefore concludes that "the isotropy of the universe and our existence are both consequences of the fact that the universe is expanding at just the critical rate."

While the anthropic principle can provide this link between a feature of the cosmos (isotropy) and our own existence, it does not, of course, provide any physical explanation of it. Such an explanation must derive from the physical processes taking place in the Big Bang — and these, at least with regard to the first 10^{-35} second, are not accessible to us. Neither by direct astronomical observation nor through simulations in laboratory experiments on Earth can we learn anything of the events in this first fraction of a second in the life of the cosmos. And the statements

113

we can make about it are therefore bound to be speculative. According to a theory by Alan H. Guth and co-workers isotropy could have arisen as a consequence of rapid cosmic inflation which might have occurred at 10^{-34} second. Rapid inflation as a kind of phase transition caused by the symmetry breaking of the fundamental forces could have smoothed out all irregularities.

THE CYCLICAL COSMOS AND THE BIOLOGICAL SELECTION OF THE NATURAL CONSTANTS

The ordered and well-understood world of the Big Bang we have described so far also has its darker, hidden side. The above description of processes in the explosive beginning of the cosmos is generally accepted without question in its version of events after the end of the first minute — that is, as far back as the synthesis of the elements; but further back than this, physics becomes increasingly uncertain and there is more and more room for alternative hypotheses. Three basic problems still await solution within the scheme of the Big Bang:
— the origin of the galaxies;
— the primordial singularity of the Big Bang itself;
— what happened before the Big Bang.
Each of these questions, in its own way, is intimately connected with the fact of our existence. Without galaxies there would almost certainly have been no life, as galaxies are the precursors of stars. In the singularity in which it began, the universe possessed definite initial conditions; and the question of why the universe began in just this state and no other is one that cosmologists have long worried about. As for the popular question often heard from non-scientists: "What came before the Big Bang?" — this is a meaningless one, given that space and time themselves came into being *in* the Big Bang. Without time, there can be no "before," just as without space there can be no "outside"!

Let us take this last question a little further. The notion of a universe "before" our own and another "after" it perpetually fascinates those attracted to the ideas of eternity and reincarnation who are eager

114

to transfer the concept of re-birth from the personal to the cosmic. The magic word is "cyclical" or "oscillating": in an oscillating universe, each Big Bang is understood to represent no more than a step along the path from an eternal past to, perhaps, an eternal future.

Despite the Hinduistic trimmings, there are a few scientific comments that can be made about this — and even some speculations. Anyone is at liberty to think of our universe as having one or more cycles (i.e. other universes) "tagged on" before it — and after it as well, if our present one should eventually collapse. But this remains an empirically untestable and therefore meaningless suggestion, as long as no information, no sort of signal, can survive from one cycle to the next. In that case, there is simply no trace of any causal connection linking these successive cycles.

Whether this situation actually exists is something no one is at present able to say. The general theory of relativity predicts a singularity for the beginning or the end of a finite universe — a state in which (at least on a purely mathematical basis) density, pressure and temperature become infinitely great. In this situation, no signal could be transported between two cycles. However, it is difficult to define "infinite" in the description of local physics. The prediction of a singularity could be wrong, for example, because quantum effects are not taken into account in Einstein's theory of gravity. And if his theory cannot be extrapolated as far as "infinity" but becomes invalid somewhere along the line where matter is still of finite density, then — in principle at least — the transfer of information to the next cycle is a possibility. There would then be a physical basis on which to imagine several cycles strung together. The Big Bang would no longer represent the beginning of the universe: as a less awesome "Big Bounce" event, stripped of its status as a singularity, it could permit information to pass through from the previous cycles to our own, and thus permit us to determine the conditions at the start of our present cycle. However, as long as we are describing the circumstances of the Big Bang on the basis of the general theory of relativity alone, we cannot decide whether or not these initial conditions trace back to the preceding cycle in any way.

115

Either way, the Big Bang, even in its guise as a "Big Bounce," would act as a sort of mincer in which everything that went before would be wholly or partly destroyed and scrambled. If we allow speculation full rein once more, what might happen in such a "rebound Big Bang"? We might expect

— the destruction of the physical conservation quantities (conservation of energy, conservation of angular momentum, etc.);
— the destruction of all particles, along with their characteristics;
— changes in the fundamental constants;
— the "biological selection" of new fundamental constants.

We can only speculate as to whether there would be such a thing as an electron, with the same or similar characteristics, if indeed such things as particles were to appear on the new cosmic stage. Would there once again be a Pauli principle, based on the premise that particles of the same type are totally indistinguishable? All electrons have the same physical identity; it is never possible to say just which electron will turn up in an experiment. "The wondrous identity of particles of the same type has never been given an acceptable explanation. This indistinguishability cannot be regarded as a triviality, but must be seen as one of the central mysteries of physics," as John A. Wheeler once remarked. One thing at least is clear: there would be no stable matter without this characteristic of the elementary particles, since this is the basis of the Pauli principle, discussed in Chapter III; and without matter there could of course be no life.

In the context of a cyclical universe it would in principle be thinkable that the fundamental constants might change in each successive, nearly infinite rebound event. Would it then still be possible to explain their numerical values — those of the dimensionless fine structure constants — on the basis of physics? Wouldn't they simply be inherent in the initial conditions, brought to life afresh in each new beginning? If the state of matter were not "singular" in the Big Bounce, that is, if temperature and pressure did not exceed some set limit, then the previous cycle of the universe would be able to affect the next. It would then conceivably be possible to draw up a complete theory of the cosmos that could deal with the conditions in a quasi-rebirth on the basis of physical laws.

Even today our understanding of the origins of galaxies is still incomplete, which means that we are still without an explanation for one of the most basic events essential to our existence as observers. There are several models of galactic formation, but all of these have unresolved details at one point or another. The variations in the density distribution of the cosmic gas in the year 300,000 PUN can only have been very minor as we know from the isotropy of the background radiation. In a *static* universe, these irregularities would have quickly intensified, and the cosmic gas would have separated into large "clouds," the potential precursors of galaxies. However, the expansion of the universe slowed down this process of galaxy formation, and the irregularities only grew slowly. What physical process brought them into being in the first moments of the Big Bang remains a mystery.

One suggested mechanism for the formation of galaxies is based on the chaotic model of the Big Bang. According to this, oscillations taking place during the dissipation of the initial chaos could have spread through the cosmic gas in the form of waves — a similar process to the spreading of waves across the surface of a pool into which a stone has been thrown. Then, as soon as radiation and matter separated and the gas became transparent, the pressure, distributed irregularly throughout space, would quickly have fallen. This would have left some areas with a high pressure and some with a low pressure. The low-pressure areas would then have imploded much as TV tubes implode. However, implosions create shock waves, which in turn cause gas to compress. This might finally have compressed some clouds to such an extent that galaxies could form.

What is better understood by now is the reason for the particular shapes of some galaxies — spiral arms, offshoots and the like. These are galaxies that have collided with others at some point in their history. (Even our galaxy might well have collided in the past.) The results of computer simulations of collisions between galaxies are given in the illustrations on pages 118 and 119.

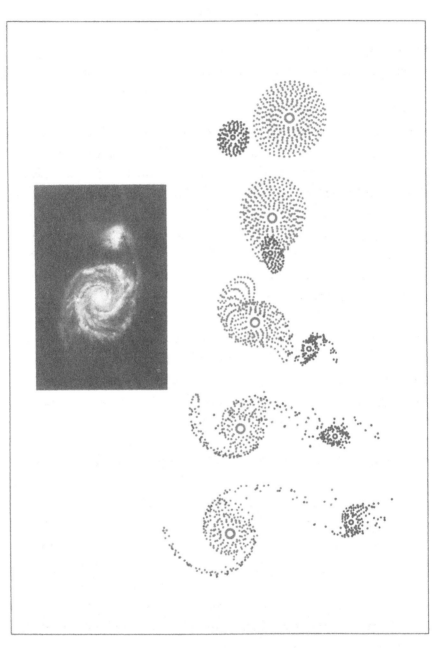

Photograph of galaxy M 51 and its companion (NGC 5195), a galaxy of smaller mass (inset). Alongside, the computer simulation of a collision between two unequal galaxies by A. and J. Toomre (1972).

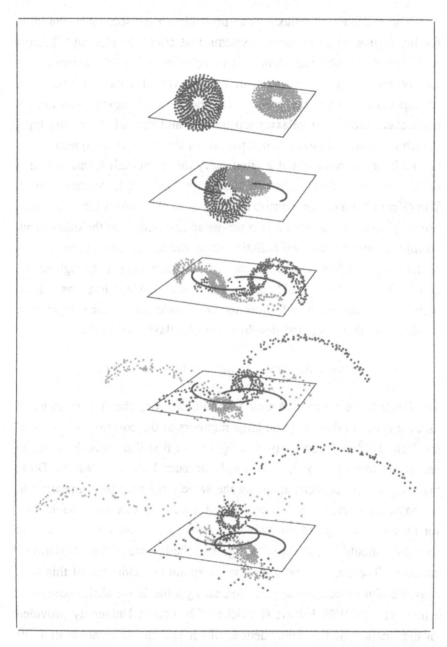

Computer simulation of two galaxies (NGC 4038 and 4039) in collision (A. and J. Toomre, 1972). A collision between two galaxies may occupy several million years; the computer can follow the process through in a matter of minutes.

The existence of galaxies was probably an indispensable condition for the formation of planetary systems that could develop into bearers of highly-developed life forms. The velocity of cosmic expansion is clearly intimately connected with the growth of galaxies: too swift an expansion hinders the process and too slow an expansion leaves it unchecked, producing galaxies too quickly and thus of the wrong type. Which fundamental laws decided the fate of the galaxies is a question that cannot be answered without a satisfactory theory of their formation. It is certain, however, that gravitation and electromagnetic force are involved. The effect of gravity on cosmic material is to slow down the expansion, bringing about condensation into pre-galactic clouds. On the other hand, cosmic magnetic fields and radiation delay the formation of these clouds, as does a gas of high-mass neutrinos distributed regularly throughout the entire cosmos. This would bring in the weak nuclear force as a third ingredient in the mixture leading to the cosmic coincidence, or cosmic inevitability, that made the development of galaxies possible.

WAS GRAVITY STRONGER IN THE PAST?

The English cosmologist Arthur S. Eddington was the first scientist of modern times to deal with the large numbers of the cosmos — figures of the form 10^{40} and powers of this figure. At first little was done but to wonder at these numbers. It was only decades later that Paul M. Dirac made public his concern at seeing the same large numbers appearing in his cosmic coincidences. He considered it quite simply improbable that, for instance, the age of the universe, expressed by means of an atomic time unit, should lead to the same figure as the inverse of the gravitational constant. There had to be a theory to explain coincidences of this sort. This led him to develop a cosmological hypothesis we shall come to in a moment. In 1961 Robert H. Dicke of Princeton University provided an explanation of this coincidence, which laid the foundation stone of the weak anthropic principle: when the age of the cosmos — the large number of 10^{40} — was different, there were simply no human beings to observe the existence of this number.

The answer to the question then of why the universe is as old as it is or as large as it is would be the same. As we said in the introduction, these arguments do not "explain" why this is the way things are; they simply demonstrate that certain cosmological preconditions must be fulfilled to make our existence possible.

Such an "anthropic" explanation was not what Dirac was looking for; he required an explanation based on *physical laws*, and his explanation of the cosmic coincidence was therefore quite different. If the "age of the universe" changed with time, this must also be applicable to the gravitational constant in order to keep the ratio of these numbers equal to 10^{40} at all times. Let us follow his reasoning. The electrical force between the electron and the proton is 10^{40} times stronger than the force of gravity between the two particles (Cosmic Coincidence No. 3; see Chapter II). The size of the universe (the Hubble "radius") is 10^{40} times the size of the proton. Two fundamental forces, then, have the same relation to each other as the linear scales of the atom and the cosmos.

We do not know why these two numbers are the same, but as the cosmic timescale is changing, this circumstance merely signifies that we are living at a special point in time. At any other point in cosmic time, the two numbers would *not* be the same. To "rectify" this, Dirac proposed that the gravitational constant also changed with time. This would mean — if it is also assumed that atomic dimensions do not change over time — the two numbers are *always* the same, and not merely at just this moment. The problem inherent in this, of course, is that of finding the laws of a general theory within which the gravitational constant is permitted to change.

By this step Dirac had opened a Pandora's box: why shouldn't other fundamental "constants" change as well — and why only in time: why not in space too? In the conventional theories of Newton and Einstein, of course, the constants are indeed quite simply *constant* numbers. Dirac's hypothesis concerning these large numbers, therefore, was quite deliberately going out on a limb and — in 1937 — did not even amount to a theory, but at first only a hypothetical cosmological model. Proper mathematical theories involving variable gravity were only put forward later: Pascual Jordan in 1955, by Carl Brans and Robert Dicke in 1961 and by Dirac himself in 1974. According to all these theories, the gravitational constant (G, or "alpha–G" in the tables in the Appendix) was stronger in the early universe. Dirac had assumed in particular that G declined in inverse proportion to cosmic time, and predicted a value of -6×10^{-11} for the relative reduction per year. It is difficult to carry out any direct observation of this effect, and attempts to do so have so far provided only an upper limit for any possible fall in G. But astrophysics offers numerous possible tests:

— With higher gravity the sun should have burned more hydrogen in its earlier phases, and thus shone more brightly than it does today;
— the orbit of the Earth would have increased in diameter with time;
— with the fall in the strength of gravitational force, the Earth should have become larger: this could have had an effect on continental drift;
— various gravitationally bound systems — star clusters, galaxies, clusters of galaxies — should have taken different forms;
— meteorites would have been heated up more strongly by a hotter sun;
— the rotation of the Earth should slow down. From measurements we know that the length of the day is indeed increasing by 2 milliseconds each century. In the process, energy is being lost to the Moon, which is moving away from the Earth in a tight spiral; a weakening gravitational attraction would encourage this. To filter out this influence specifically, the more major effect of tidal breaking must first be deducted.

The observations we have at present cannot entirely exclude the possibility of a fall in the strength of gravity with time, but indicate that it cannot be diminishing by more than 10^{-11} or 10^{-10} of its value per year. Dirac's predication is squarely in the middle of this range of permissible maxima. We return to the precise effects on the sun and the Earth of a change in gravity in Chapters VI and VII. A fall in the strength of gravity at such a rate as Dirac envisaged is most clearly excluded by geochemical data; but a less rapid trend is nevertheless conceivable. A change in gravity would have consequences for the stars, and a reduction in brightness not only of the sun, but of all other stars would mean that all galaxies would shine more weakly in the course of time. Besides this, the distance from each star to the center of its galaxy would increase, and our galaxy, along with all others, would slowly expand. Pulsars — rotating neutron stars — would turn more slowly, in the same way as the Earth. The rotation of a pulsar provides us with a highly precise clock, accurate to one billionth of a second; this pulsar clock would gradually "tick" more slowly if gravity decreased and the consequence would be an increase in the number of slowly rotating pulsars. To measure this change over time, we would of course have to use clocks that were not driven by gravity, and atomic clocks are ideal for this purpose: their recording of time would remain constant whatever the changes in gravity.

Dirac's hypothesis could be directly tested if such an atomic clock were to be set down on Mars' moon, Phobos. If gravity is decreasing, the orbit of Phobos around Mars should be getting larger; if the clock and the orbit of Phobos were to be observed together, the change in G could be directly measured.

The first firm value for a change in G over time, rather than an upper limit derived from experiments, was sprung on the scientific world in 1975 by the Dutch astronomer Thomas van Flandern. After comparing the motion of the moon and the rotation of the Earth with the gravitationally independent atomic clock, he claimed that the gravitational constant was reducing by 8×10^{-11} of its value per year. With this announcment, van Flandern had clearly gone too far for his fellow scientists who tore him virtually limb from limb. The criticism was particularly directed against the complicated and therefore potentially error-prone method by which he evaluated his observations. Van Flandern's suggestion has remained under suspicion since 1975, but his claim has been slowly gaining acceptance. Similar research by other investigators has tended to confirm his results, and the least that can be said is that it has yet to receive the disproof that the outcry might lead one to expect.

Indirect conclusions regarding a change in gravity can be drawn from the production of elements in the Big Bang: the amounts of cosmic deuterium and helium are very sensitive to the strength of gravitation at any given time. But a minimal decline in gravity is not yet ruled out with certainty by this consideration.

How Universal are the Fundamental Constants?
—Observed Limits of Change

What kinds of changes to the laws of nature can we imagine in the first place (see Chapter I)? The easiest thing is to assume that any change will take place in a continuous manner in space and time (though, of course, a change to the laws of nature carried through in "jumps" would also be conceivable.)

If — as a tenet of some "super-theory," for instance — the familiar laws of nature are to change in a continuous way, the simplest possibility is for the fundamental constants appearing in these laws to change. Up to a point, this concept can at least be tested by experiment.

And a test for such heretical hypotheses is badly needed, for a dependence of the laws of nature on space and time would radically affect astronomers' understanding of their field. Everything we know about physical conditions applying to astronomical objects is deduced from observations on the assumption that the object's natural laws, along with their constants, are exactly the same as here on Earth. This applies to the exploration of Mars, Venus, Jupiter, and the sun, all the way to the most distant systems, quasars and the Big Bang itself.

If the laws of nature did indeed differ from place to place and from point to point in time, we would only be able to observe the universe in an oddly distorted way. Tied as we are to the Earth, where our telescopes are sited, we would have to apply corrections to all of our observations to take into account the nature of the variation.

There are two groups of dimensionless fundamental constants: the fine structure constants of the four forces of nature; and all the independent quotients of the masses of the elementary particles (see Chapter II).

Changing the fundamental constants is a complex game to play. Of course, any of the individual constants can easily be "tweaked" to a new value. But then the experimenter will have to comb through all of physics (or think up new experiments) to see whether the new value contradicts any observations! The gravitational constant was discussed in the last section; but other possibilities have also been examined — most thoroughly by Freeman Dyson in Princeton and by W. Eichendorf and Michael Reinhardt in Bochum, Germany. In short, their studies have shown that the prospects of a possible time-dependency for the fundamental constants are not good.

The electromagnetic fine structure constant, alpha–E, whose variability was first considered by George Gamow, shows a high degree of constancy over time and space — achieving a tolerance of one part in 10,000 over the last billion years. This has been demonstrated by observations of spectral lines in the most distant quasars, which did not fit with Gamow's idea. If these results from optical astronomy are combined with radio measurements, they also strongly suggest that the *relation* of electron mass to proton mass (= 1/1837) has remained constant over cosmic periods of time (see Chapter V).

Precise measurements in laboratories on Earth can establish whether the electromagnetic fine structure constant and the mass relationship of electron and proton change depending on their location in *space*. These quantities largely determine the construction of atoms and molecules, so this could provide an indirect route for the study of possible changes. Spatial variations would be expressed in a place-dependent total energy for each material, according to its chemical make-up. One result of this would be that different bodies would fall at different speeds in the gravitational field of the Earth. Fall experiments — whose prototypes were carried out by Galileo Galilei (as legend goes) at the leaning tower of Pisa — could exclude such a variation with a high degree of precision.

Another natural constant is the value of an elementary electrical charge. We know from the distribution of isotopes in chemical elements on Earth that the electrical charge can barely have changed during the planet's history. There would have been a noticeable influence on electrical repulsion between atomic nuclei and as a result on the process of radioactivity, and altered radioactivity in some isotopes would have changed the mixture of isotopes according to age. However, it has not been possible to trace this effect: an analysis of old rock has been shown by Dyson to limit the possible variation in the elementary electric charge to an annual change of less than a ten-thousand-billionth (10^{-13}) of its value.

A further consideration is the weight (mass) of all the elementary particles. Did they change in the course of cosmic history? Speculations about such a variation in the masses of the elementary particles over time were introduced in 1974 by the American physicist S. Malin, who offered the choice of either an increase or decrease. We can exclude a decrease without further ado. But an increase? Here the history of the Earth once again comes to our aid. Three-and-a-half billion years ago, when life was first appearing on Earth, an increase in mass would have resulted in a terrestrial temperature of $-200°$. And even seven hundred million years ago, when coral formations were being deposited, the Earth's surface would instead have been below freezing (see Chapter VII).

Many speculations on the variability of the laws of nature can thus be shown to be limited in their variability, to say the least. But the consequences to be drawn from the anthropic principle reach even further back in time and out into space.

The "Big Crunch"

Although a closed model of the universe has yet to find favor with astronomers, this cannot as yet represent a final verdict. Too often, the interpretations of astronomical observations of distant galaxies and quasars have proved in need of revision in the light of new findings. A finite universe — one that expands and in the course of time collapses again — must be given consideration; to close this chapter, we shall examine it in the context of the anthropic principle.

A finite cosmos must exist long enough to permit the existence of life for at least part of its existence — that is, for at least a few dozen billion years. On the other hand, it could be that the closed universe is

126

already approaching its full life expectancy. (If our universe is a closed one, we are living around the end of the first third.) This would not only make the position of the human race an especially privileged one, as suggested by Dicke's Weak Anthropic Principle; it would go further and set considerable limits on a fundamental property of the universe, in this case its life expectancy.

Two questions are of particular interest to us here. What happens as the universe shrinks toward collapse? And how would a life form be affected by its collapsing environment? At first the expansion would continue for a while, before coming to a halt; in the subsequent contraction phase, first the galactic clusters then the galaxies themselves would close in on each other. How will the cosmos change after the moment at which expansion becomes contraction? We could expect our neighboring galaxies to move toward us first, their light slowly shifting from the red to the blue end of the spectrum. More distant galaxies would continue moving away from us for some time, with red-shifted light; the light from far distant objects would still be reaching us from the previous expansion stage. Clusters of galaxies, which now occupy about one hundredth of the entire volume of space, would merge with each other when the cosmos had shrunk to 20 percent of its maximum expansion. Individual galaxies, and the stars in them, would move with increasing speed, like molecules in a gas undergoing continuous heating.

Although stars would then collide more frequently, what would "upset" them far more would be the rising temperature of the cosmic background radiation. From its present value of 3 degrees Kelvin it would rise to several thousand degrees once the cosmos had shrunk to a thousandth of its present volume. At this point, shortly before the "Big Crunch," the night sky would be as bright and as hot as the surfaces of the stars themselves. After that, the Big Crunch would be like fireworks within a glowing hell — a cosmic replay of the "primeval soup."

All life, by this time, would have met its end by roasting and vaporizing; but some thought has been given to the question of whether a sufficiently intelligent life form might not, subjectively, be able to enjoy an infinitely long life with the finite duration of a closed cosmos. The idea, which comes from Frank Tipler (Tulane University, New Orleans), runs like this: Consider the phenomenon "life" in the abstract, as a system that collects, stores and transfers information. Then, every process by which an information unit — a "bit" — is exchanged represents an elementary part of the life process, and requires a certain measure of time. If the tempo of the metabolism of a life form is increased, the individuals — subjectively, at least — would live longer. They would be able to carry out more elementary life processes per second. Is it conceivable, Tipler wondered, that a life form might so increase the speed of its metabolic processes during the collapse towards the Big Crunch that, as cosmic time runs out, it can get through more and more elementary life processes, finally achieving an infinite number before the physical conditions of an ever hotter cosmic hell-fire destroy all possible organization?

Speculations like these should not be brought to heel too quickly by hidebound considerations of plausibility — for who knows, after all, what shapes might eventually be taken by a highly developed species? But as long as biological systems play a fundamental role in their make-up — as they do in terrestrial life — a few doubts may be given vent to. There is no way that biological life can be endlessly accelerated. Certain reactions within energy and information processing — such as eating, learning, and exchange of knowledge — are tied to a minimum time requirement, and here biology sets a limit to any such life form. "The biological clocks," says Dyson, "can never speed up fast enough to squeeze an infinite subjective time into a finite universe." This situation, of course, is fundamentally different in an everlasting infinite universe.

Chapter V
What Quasars Tell Us
About the Laws of Nature

How the World Began

As any astronomer will tell you, if you want to know about the past you need only look a long way off. Since it takes light a full second to travel 300,000 kilometers, it is often quite some time before the light from a distant source travels the length of the cosmos and reaches a terrestrial telescope. This means that the light from the more remote sources comes to us from an earlier cosmic epoch. The most distant and thus oldest object of astronomical observation cannot actually be localized — it is an "event" and happened everywhere: the Big Bang. The waves that reach us from this earliest event have been familiar since 1965 as cosmic background radiation. They have been on their way to us for about twenty billion years.

But number two on the list of the most faraway things in the universe is a class of material objects, which look just like ordinary stars when they appear on photographs of the night sky, and were long thought to be nothing more. It was only after their high-powered radio emissions were discovered that they became known as "Quasi-Stellar Radio Sources," or *quasars* for short. They are several billion light-years away from us, and as a result of the expansion of the universe their speed of recession exceeds that of all known galaxies. The most distant quasar is receding from us at over 90 percent of the speed of light, fleeing with the very edge of the universe and sending us light from the "beginning of the world."

Since distant objects are as old as the time taken by their light to reach Earth, we can use them directly as a source of information on the nature of physics in that place and at that time. We can compare this information with the state of physics in our laboratory on Earth and draw conclusions from it. This is astronomy in the form of astro-archaeology — but not the way Erich Van Daniken might do it: if we examine the laws of nature prevailing in distant objects, this is comparable to the work of terrestrial archaeologists, who can deduce conditions on Earth millions and billions of years ago by examining rock formations and the yearly growth-rings of fossilized trees. In astronomy, the information we seek is stored in the light waves sent across space by a star. As time passes while these waves are on their way, remoteness in space and remoteness in time for the astro-archaeologist amount to much the same thing.

The purpose of the anthropic principle is to establish how closely the existence of life is bound up with the structure of the universe, and in particular to what degree the laws of nature, represented by the fundamental constants, could have been different; so it is especially important to know whether the fundamental constants had different values at some time in the past, and if so, how their relationships have changed since. The light from far-distant astronomical objects allows us to read the long-past values of some of these constants, among them those which determined the characteristics of the light waves sent out by quasars:

— The light waves generated by the electrons in the shells of chemical elements enable us to deduce the electromagnetic fine structure constant (alpha–E);

— The radio waves emitted by the nuclei of hydrogen atoms — that is, protons — proved a figure dependent on alpha–E, the strong nuclear force (alpha–S) and the mass relationship of electron and proton (m_e/m_p).

To achieve this, light waves are compared with radio waves, with particular reference to the 21-centimeter radio emission of the hydrogen atom, a spectral line produced when the proton of the hydrogen atom reverses its "spin." It does this on average several billion times per second, thus producing radio waves in the gigahertz region. (A wave

1.4 gigahertz in frequency has a wavelength of 21 centimeters.) This frequency is a physical consequence of the mass of the proton that is constantly "switching" from one spin state to the other and back. Optical quasar lines, on the other hand, owing their existence to the activity of the electrons of certain atoms, are influenced by the mass of the electron. A comparison of light waves with radio waves thus provide a value for the mass relationship m_e/m_p at the time the radiation was sent out.

The idea of testing the natural constants by means of the light from astronomical sources had already been suggested by the physicist M. P. Savedoff of the University of Rochester seven years before the discovery of quasars, in a small note in the scientific journal *Nature*. At that time, the only related light and radio observations available to him were those made of a large galaxy in the constellation of the Swan. A light source had just been identified for the first time, and the source in question was the galaxy Cygnus A. Compared with the quasars, this galaxy is relatively close to us, only some three hundred million light-years away — about 150 times as far as our nearest neighbor galaxy, the Andromeda nebula.

Savedoff's results had already pointed to a strong likelihood that the constants alpha–E and m_e/m_p in Cygnus A and on the Earth — that is, in the galaxy Cygnus A three hundred million years ago and on Earth now — were identical at least to a few parts in a thousand. This implied that, within the region including Earth and Cygnus A and within the period of time concerned, the constants had changed by at most that amount.

That was a start. Could this investigation, so important to an understanding of the depths of space, be expanded to even larger stretches of space and time? It was only by using quasars that the boundaries of knowledge could be pushed as far as the edge of the visible universe. Let us go back to that memorable day in February 1963.

THE STORY OF QUASARS

Illumination came to the Californian astronomer Maarten Schmidt on the evening of February 5, 1963. "That evening," he relates, "I went home assailed by doubts. Something incredible happened to me today,

131

I told my wife." What had caused Schmidt so much astonishment was the behavior of a mysterious object recently discovered in the sky; in the Third Catalogue of Radio Sources compiled at Cambridge University, 3C, it was entered as the strong source 3C 273. In December 1962, Schmidt had photographed the light of the star identified with this source and discovered six wide, bright lines in its spectrum that did not seem to match any of the familiar spectral lines of atoms such as hydrogen. On February 5, Schmidt finally achieved a breakthrough towards understanding the puzzle: the lines exactly matched those of hydrogen (and in one case a line belonging to ionized magnesium) if he assumed an unusually large red-shift of the spectrum — an indication that object 3C 273 was moving away from the Earth at about 15 percent the speed of light! In making this assumption, Schmidt had identified the first *quasar* — although the name was not given to these strange radio stars by the astrophysicist H. Chiu until 1964.

Today we know that all of the quasars so far identified show a different but large red-shift of their spectral lines, and often radiate few radio waves, or even none at all. However, the name "quasar" has prevailed.

Considering the very short history of research into these objects, Schmidt put his finger on the essential point very quickly. The catalogues had for a long time included a number of strong radio sources for areas of the heavens where, unlike most radio sources, there was no galaxy on the photographic plate, but at most a handful of weak stars. Fishing the right star out of this group — in the days when radio telescopes supplied notoriously vague coordinates — was not an easy matter, and a way of doing this with the help of the moon was first arrived at in 1962 by Australian radio astronomers. When the moon passed in front of the source and hid it from view, it was possible to calculate the precise position of the moon on its orbit at the moment when the source disappeared, and so arrive at exact coordinates. In the case of 3C 273, the brightest of the suspect stars turned out to be the counterpart of the quasar; it has remained the brightest known light source among quasars up to the present day.

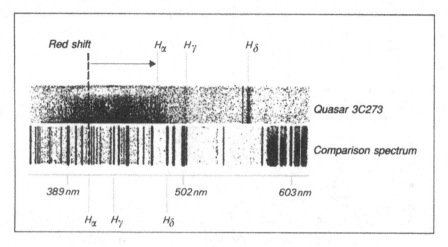

The red shift in the light from Quasar 3C 273 reveals itself when its spectrum is compared to a spectrum in which the lines of hydrogen are unshifted. The red shift of z = 0.158 corresponds to a recessional velocity of 45,000 kilometers per second for the quasar.

As harmlessly as the quasar story began, the more astronomers learned about them, the harder the nuts the desperate theoreticians had to crack. Only decades after Schmidt's first breakthrough, we suppose we might understand at least the fundamental mechanism that drives quasars and other energy-rich objects like them, based on a model constructed around super-massive black holes. At the same time the number of unanswered questions central to the construction of this model has continued to grow.

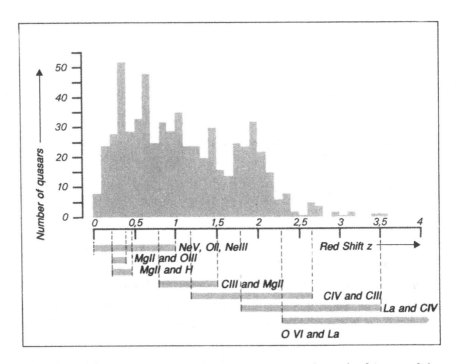

The over 600 known quasars transmit information from the fringes of the visible universe. From their light it can be ascertained whether several of the natural laws known to us deviate over great distance and in the cosmic past. In proportion to the red shift, various lines are used for its determination: magnesium (Mg), oxygen (O), neon (Ne), carbon (C), lanthanum (La). Past a red shift of z = 2.2 the number of quasars falls off rapidly.

What made quasars so mysterious right from the start was their high luminosity — which originated, moreover, in an extremely small region of space. Quasar 3C 273, for example, shines five million million times more fiercely than our sun, achieving the same amount of radiation as an average galaxy, which will typically consist — like our own spiral galaxy — of 100 billion individual stars. To this must be added the finding that quasars often vary in brightness over a period of only a few days, which indicates that their entire radiation must come from a region with a diameter of only a few light-days, hardly larger than the solar system. Radiation frequently many times as strong as that of an entire galaxy must be originating in a region with only a billionth of its size.

Almost as Fast as Light

"A successful model must explain an engine which may be the most energetic in the universe and at the same time must force this engine into a container of minute dimensions." This was Martin Rees' description of the quasar dilemma.

The red-shift of the spectral emission lines is usually explained by the Doppler effect: the light from a star is received on Earth shifted to a longer wavelength if the star is moving away from us. The difference between the wavelength emitted and the wavelength received — the red-shift— is thus a measure of the relative velocity of recession. (If the star is moving towards the Earth, there is a corresponding blue-shift.) Schmidt's quasar 3C 273 had a red-shift value of $z = 0.158$, equivalent to 15 percent of the speed of light. Since then, quasars have been found that are receding at a velocity of over 90 percent of that of light.

If this interpretation of the red-shift is correct, quasars are seemingly the fastest astronomical objects we know. But how do quasars achieve these enormous velocities? This phenomenon is explained on the assumption that this movement is caused by the expansion of the universe, and this assumption has therefore been called "the cosmological hypothesis." We can now derive from it a statement regarding the distance of the quasars from us.

The estimate of the distance from Earth to the quasars depends on the assumption that the universe is expanding (equally in all directions). We observe this expansion in the movement of distant galaxies away from us and each other. Hubble demonstrated in the twenties that the recessional velocity of the galaxies rises with increasing distance. The red-shift of the spectral lines permits deductions to be made concerning the regularity of recessional velocity and, in turn, the distances involved. Given their great distance from us, however, quasars must possess the immense luminosity we mentioned in order that we can see them at all.

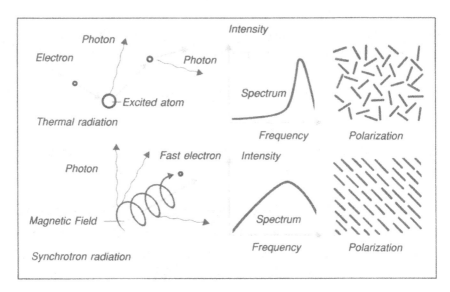

Electromagnetic waves of a specific type are given off by electrons when they are accelerated by a magnetic field: synchrotron radiation. The electrons move in spirals because of it along the lines of the magnetic field. The waves are then preferentially directed (polarized) perpendicular to the field. The spectrum also has a characteristic form which differs from the case where electron movement is disordered, as in the hot gas of a star. In that case, thermal radiation is formed with disordered (non-polarized) waves.

Many astronomers found this difficult to swallow and therefore looked for other possibilities, doubting the validity of the cosmological hypothesis. An alternative possibility does exist: the red-shift might be caused not by relative motion but by gravitational fields within the quasars. Light which has to battle against a strong gravitational field to reach an observer loses energy, which likewise brings about a red-shift. If the cosmological hypothesis is abandoned, the quasars are no longer necessarily the most distant visible objects in the universe: they could even — as the "local hypothesis" proposes — belong to our own galaxy. As a consequence, their luminosity could be a good deal lower. The main obstacle to this interpretation lies in the problem of coming up with plausible astrophysical models for quasars on the basis of the local hypothesis. No attempts in this direction have succeeded as yet. For this reason, although there are occasional minor controversies, the interpretation of the red-shift based on the cosmological hypothesis is accepted by most astrophysicists.

It was confirmed when observations showed that quasars are located within galaxies, and that the stars of these galaxies possess the same red-shift as their respective quasars. It is hardly suprising that such observations were difficult to make for these distant objects: a quasar at that cosmic distance is so bright that it totally obscures weak stars in its vicinity.

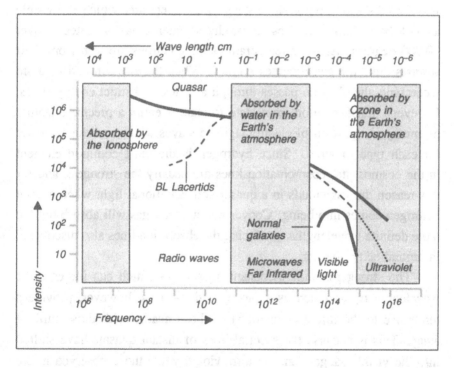

Normal galaxies like the Milky Way give off thermal radiation which is generally made up of the light from their individual stars. The synchrotron radiation of quasars and their related BL lacertids is very different. They exceed the brightness of normal galaxies not only in the UV-range, but also rise in the radio-range as well.

For some quasars, with a relatively small red-shift — plus, in 1971, quasar 3C 273 — has it been possible to make out galaxies or galactic clusters with matching red-shifts in the immediate vicinity. At least in the case of these nearby quasars, the red-shift is established as a measure of their distance from us.

It is believed that indications of the nature of the physical processes taking place in the interiors of quasars can be deduced from the characteristics of the emission lines by which the red-shift is established. There are two sorts of spectral lines: emission lines, radiated by atoms and molecules at quite specific wavelengths and appearing as sharp bright lines in the spectrum; and absorption lines, which arise in the opposite way from the effect of atoms and molecules filtering out particular wavelengths and leaving sharp gaps in the spectrum, which are visible as dark bars. Emission lines are produced when a gas is heated to over 10,000 degrees. At these temperatures the electrons are torn from their atoms so that the gas becomes ionized. When the electron and its atom recombine, the electron passes through a series of distinct energy states. On every transformation from state to state it emits a precise quantum of energy in the form of electromagnetic waves which are characteristic for each type of atom. Since hydrogen is the most common element in the cosmos, its recombination lines are usually the strongest, and for this reason the gas clouds in a quasar emit additional light when neutral hydrogen comes into being. Conversely, a cooler gas will absorb light at quite definite wavelengths, producing the absorption lines also discovered in around sixty quasars.

The strong red-shift of distant quasars — which can increase the wavelength by as much as a factor of 4 — does, however, provide a hindrance to the direct comparison of those nearby with those furthest away. This is because the spectral lines of distant quasars have shifted into the visible range from the ultra-violet, while those observed in the nearer ones are actually produced at visible wavelengths. Strictly, then, we are seeing radiation resulting from different emission processes in each case. This gave rise to the question: were nearby and distant quasars actually the same sort of object?

Early in 1977, an observation of the veteran (nearby) quasar 3C 273 threw some light on this. The largest telescope ever launched with a spacecraft was aimed at 3C 273 from a high altitude rocket by a three man team of astronomers at Johns Hopkins University who recorded its ultra-violet light for four minutes. This enabled the researchers to "see" emission lines that otherwise only show up shifted into visible wavelengths in the radiation of the most distant (and thus fastest-receding) quasars — and even then only just. The recorded UV spectrum was found to match the sharply red-shifted one of the far-distant quasars — indicating that nearby and distant quasars are the same type of object. But where do quasars draw their immense energy? Although this question is still a source of controversy, the conviction is beginning to prevail that thermonuclear sources are totally insufficient and the driving force must lie in strong gravitational fields. Taken overall, quasars are now seen to be bodies which

— are often identified first as radio sources, although they appear as ordinary stars in the visible range;
— can show variations in brightness within the space of weeks or days;
— emit strong ultra-violet radiation;
— demonstrate broad emission lines (and sometimes also absorption lines) in their spectra; and
— manifest sharply red-shifted spectra.

A SOURCE OF SUPPLY FOR QUASAR ENERGY:
SUPER-MASSIVE BLACK HOLES

Theoreticians have been attempting to reduce this confusing array of individual observations to a single closed model since 1963, with varying degrees of success. The conclusion that quasars are not appreciably different from phenomena taking place in the central regions of active galaxies appears to be gaining general acceptance. This theory has beat out some earlier bizarre ideas on the subject — such as that quasars result from the breakdown of matter through irradiation by anti-matter, or even that they should be interpreted as evidence of extraterrestrial civilizations. Most present-day quasar models start by assuming large amounts of matter of various types, with between one million and one billion solar masses, concentrated in a very small region of space, at most one or two light-years in diameter. In speaking of "various types," the theoreticians have three main possibilities in mind:

— dense clusters of stars;

— fast-rotating giant stars, known as "spinars" or "magnetoids," and

— matter collapsing into super-massive black holes.

In dense star clusters, it would be supernova explosions above all that would produce the enormous amounts of energy required by quasars, giving the effect of lights being switched on and off like the lights on a Christmas tree. The escape of energy from rotating stellar giants with strong magnetic fields would be limited by the possibility of their becoming unstable — as a result of their high speed of rotation, for example — and breaking up. The greatest radiation output, then, is expected from matter falling into one or more black holes.

A black hole will typically come into being as the result of a star's having burned up all of its reserves of fuel in nuclear fusion. If the mass of the star is great enough, the star will then collapse through the force of its own gravity. The collapsed remains of this star will then be surrounded by a region of space in which the gravitational field is so strong that not even light can escape from it. In the case of a star of one solar mass, this region would have a diameter of 3 kilometers.

The formation of a *super-massive* black hole takes place somewhat differently. An entire star cluster of several million solar masses, along with the gaseous remains of supernova explosions, shrinks as a result of its own mutual attraction and friction and radiation phenomena into an ever smaller space, finally falling in on itself in a gravitational collapse. A black hole of 10^6 solar masses formed in this way would have an effect over about four times the diameter of the sun. The center of a galaxy is the natural location for such objects to come into being.

Even this bare outline of the course of events, however, indicates that the accumulation of mass at the core of any galaxy, deriving from dramatic processes of this nature, will nearly always permit the formation

How the center of a galaxy can develop.

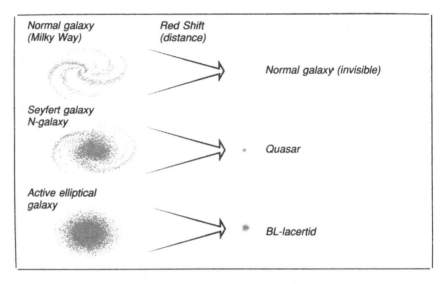

Normal galaxy
(Milky Way)

Red Shift
(distance)

Normal galaxy (invisible)

Seyfert galaxy
N-galaxy

Quasar

Active elliptical
galaxy

BL-lacertid

Quasars (and their related BL lacertids) are possibly early stages of different galaxy types. A normal galaxy would no longer be visible at a quasar's distance.

of a super-massive black hole as the natural outcome of galactic evolution. The models first in the list above may then represent, as an intermediate stage in this development, the birth of a quasar. But it is only in its black hole stage that it reaches full bloom and becomes transformed into the most efficient energy converter in the known universe. The black-hole quasar model is now supported by astrophysicists in Cambridge (England), Pasadena and the Max Planck Institute for Physics and Astrophysics in Munich. The first to draw attention to this possibility was the physicist Donald Lynden-Bell in Cambridge, in 1969.

In this model, the observed radiation from quasars arises from the process of gas and stars falling into the black hole from the galaxy, possibly colliding with other stars during their fall or even breaking up through the tidal effects of the gravitational field. The black hole takes on the aspect of a gigantic plug-hole, collecting material, accelerated to higher and higher velocities and heated by compression and friction from a disc rotating around the hole. The black hole is by this time rotating about an axis that is aligned perpendicular to this disc and thus most probably perpendicular to the plane of symmetry of the galaxy. Strong magnetic fields in the disc, enhanced by turbulent motions in the gas in concert with shock waves surging through it, cause highly relativistic electrons to convert the energy of their fall to synchronous radiation.

"Fountains" and "Twin Exhausts"

If the amount of gas flowing into the quasar exceeds a critical value, it can happen that "fountains" of fast particles are flung out of the collapse zone. Speculations on such a model are currently occupying Munich astrophysicists, among others. The supposition is that, although most of the gas catapulted out of the quasar would be halted by surrounding clouds, some rays of gas would leave the black hole along the rotation axis of the galactic nucleus — a process which would provide an explanation for the binary radio sources still perplexing atronomers.

If the "appetite" of a black hole of 10^6 solar masses is satisfied with an intake of only a few solar masses per year, it will nevertheless be able to emit the energy typical to quasars of 10^{46} erg per second — 10^{13} times the luminosity of the sun. The emission lines can also be well accommodated by the model: from the calculations carried out by researchers who have constructed disc models, it appears that the gas both inside and outside the disc can be heated up under the conditions prevailing to over 10,000 degrees; strong radiation can then drive the hot gas clouds outwards at supersonic velocities, so the characteristic emission lines will be produced at the exterior of the quasar. The radiant gas clouds will cover an area of at most a few light-weeks in diameter, and variations in the clouds provide an explanation for the observed variations in brightness of quasars. The absorption lines have proved up to now to be more difficult to deal with; in some cases they appear to have their origin in the quasar itself — a hypothesis that has been examined by the astrophysicists Judith Perry and Rudolf Kippenhahn at Munich's Max Planck Institute, but in other cases the red-shift of the absorption lines differs so greatly from that of the emission lines that a better explanation may lie in cooler gas in clouds or galaxies between the quasar and the Earth, which may filter out absorption lines from the light of the quasar behind them.

It is also interesting to consider what happens to quasars after their "death" — their activity is generally supposed to come to an end after some ten to one hundred million years. As long ago as 1969, Lynden-Bell was already drawing up a scenario for their death that would leave things relatively quiet around the super-massive black holes in galactic centers after the active quasar phase. On an estimate by Lynden-Bell and his colleague Martin Rees in Cambridge there are probably at present a hundred thousand times more "dead" than "living" quasars. They believe that there is a now silent quasar in nearly every large galaxy. Their weaker but still observable activity derives from a continuing but sharply reduced inflow of galactic gas.

These earlier suppositions have since received support from some astronomical observations. In recent years astronomers have found indications of a central black hole in at least three galaxies:

— In 1976 a strong localized X-ray emission was discovered in the center of the nearest radio galaxy, Centaurus A, indicating a super-massive black hole of about one hundred million solar masses.

— In 1977 a group of researchers at the University of Pennsylvania using the National Radio Astronomy Observatory found a compact radio source that was almost a point source in the center of our own galaxy. Since then, radio and infrared observations indicate the existence of an accretion disc at the core of the Milky Way — which astronomers would expect if a black hole were present. The estimated mass of this putative black hole is six million times that of the sun.

— In 1978 Californian astronomers spotted a new feature in the region of the nucleus of the elliptical galaxy M87, which is about fifty million light-years away and has long been notable for its spectacular bursts of activity. Observations suggest that its stars are more densely concentrated about the nucleus than is usual and show abnormal velocities.

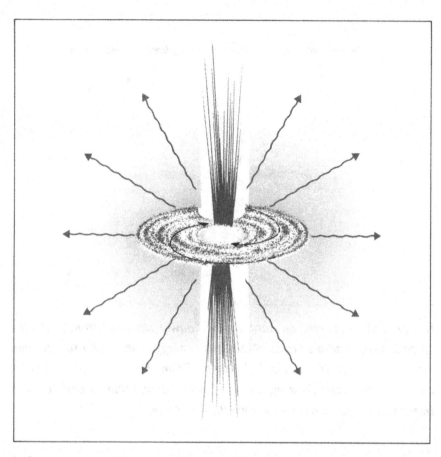

A fountain resembles one of the models which astrophysicists want to use to explain the phenomenon of quasars. According to it, a rotating disk of gas and stars pour matter into a super-massive black hole. As in fountains, particles are ejected from the area of collapse.

DOUBLE AND TRIPLE QUASARS

In 1979, quasars once again made a spectacular appearance on the astronomical stage when astronomers came upon a pair of quasar twins. According to present ideas on the subject, these could be the double image of a single quasar. The current hypothesis suggests that the apparent duplication of the light from the quasar is affected by a "lens" that is itself as large as an entire galaxy, and which must lie somewhere between the quasar and the Earth.

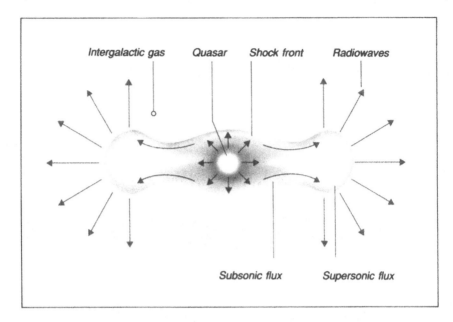

It is possible that some quasars are the "power stations" from which the so-called twin radio sources obtain their energy. Such radio sources can extend as far as 10 million lightyears. From the center, gas could be ejected in two directions by a kind of twin exhaust until it collides with intergalactic gas. Radio waves would then form.

The two quasars 0957 + 61 A and B were discovered six arc-seconds apart on March 29, 1979 with the 2.1 meter telescope at the Kitt Peak Observatory in Arizona. They showed identical redshifts and spectra. (The "double quasar" is actually a triple system, but the third image is very faint.) The red-shift of the cosmic twins corresponds to a recessional velocity of 212,000 km/s (about two-thirds the velocity of light). This gives a distance of around ten billion light-years, putting the object at the more distant realms of space. Our own solar system was just coming into being when the light we are observing had covered half of the distance from its source to us. Compared with this, the distance between the "two" quasars is extremely small: only 220,000 light-years, twice the diameter of our own galaxy.

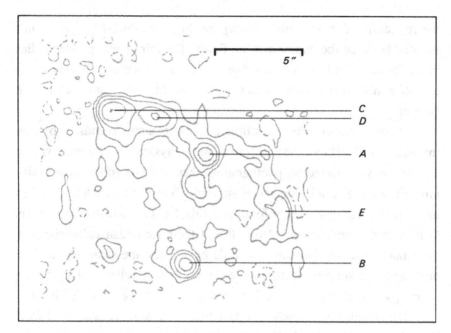

Highly precise measurement of the radio emissions from the double quasar 0957 + 561 A, B in the constellation Ursa Major, made possible with the help of the VLBI technique. A "gravitational lens" of galactic size probably creates images A and B as reflections of a single quasar. The secondary centers C,D, and E can form through irregularities in the "lens."

Speculation that the two close companions could be the twin images of a single object began as soon as they were discovered. The effect could be produced by a galaxy-sized mass lying between the quasar and the Earth, which would deflect the light and duplicate it by a "gravitational lens" effect. Such an effect had already been predicted in 1964 by the Norwegian astronomer Sjur Refsdal at the Hamburg Observatory. The justification for this idea lies in the fact that the components A and B are too alike in their optical characteristics to be considered as two separate objects: the red-shift observed in the emission lines of certain ions in the radiation source is identical in both. The quasar twins are also alike in the red-shifts of their various absorption lines, produced by gas clouds situated between the quasar and the Earth. The third correspondence lies in the fountains of material hurled out by the quasars themselves as much as half a million light-years into space: these have the same velocity in both cases.

These features were confirmed almost simultaneously by astronomers at the Harvard-Smithsonian Astrophysical Observatory, whose deductions were based on photographs taken with Arizona's new multi-mirror telescope. Such a degree of similarity between two celestial bodies had never been observed before; and this led the British astronomers Dennis Walsh and Robert Carswell and their American colleague Ray Weyman to ask whether the two could be one and the same object. The duplication could have come about through the gravitational field of a nearer galaxy, which would still be too far away to be observed itself.

The possibility of such an effect had been demonstrated by Albert Einstein as long ago as 1936. If a point light source lies directly behind the "gravitational lens" it will appear as a bright ring; if it is moved to one side of the Earth-lens line, two objects of different brightness will become visible. Thus, Quasar B's greater brightness can also be explained by the lens effect. Furthermore, the light producing the two images will have arrived by routes of different lengths, and this will produce a delay of several months between the two: variations in the brightness of object A should be visible in object B some months later.

Quasars and Changes in the Natural Forces

With the help of quasars the question of the overall mutability of the natural constants can be approached in a particularly fruitful way. If (as is always assumed) the red-shift of spectral lines in the light from quasars is caused by cosmic expansion, the relation between two wavelengths observed in the quasar's spectrum should be just the same as on the Earth. However, this only holds good as long as the constants have not changed in the meantime.

When an atom absorbs or emits energy in the form of light, an electron hops from one permitted orbit in the shell of the atom to another. In the case of spectral lines, which owe their origin to transitions between adjacent electrons orbits, the wavelength is dependent on the electromagnetic fine structure constant. In fact, the term "fine structure constant" derives from the "fine structure" in the spectrum, that is, the existence of closely spaced line radiation. An analogous process in the atomic nucleus is the 21 centimeter radiation of hydrogen. The nucleus of an atom can also absorb or emit energy by a sudden change in its orientation relative to the atomic shell. The physics of this process depends on both the proton mass and the strong nuclear force. The radio waves emitted at a wavelength of 21-centimeters are in themselves an important aid in studying the distribution of hydrogen throughout the cosmos, since hydrogen is the most common cosmic element.

The relationship of the 21-centimeter radiation to any other line in the spectrum is dependent on

— the electromagnetic fine structure constant;

— the strong nuclear force and

— the relationship of electron and proton masses.

During the seventies, American astronomers examined the radio and visible radiation from quasars in various different directions. In every case it could be shown that the quotient of the two wavelengths matched the terrestrial laboratory values to within one-tenth of one part per thousand. As it is unlikely that all three values changed in just such a way that all resulting changes cancelled each other out, each of the constants must have remained unchanged across the distance in space from the quasar to us as well as throughout the time taken for the light to reach us. For the quasars that have been examined, this amounts to about thirteen billion light-years in distance and about thirteen billion years in the past. At this epoch in the history of the cosmos, the first stars were just coming into being. The sun did not form until around five billion years ago.

With a little elegant reasoning, more far-reaching consequences can be drawn from these measurements. In March 1980 it occurred to A. D. Tubbs of the National Radio Astronomy Observatory (Charlottesville) and A. M. Wolfe of the University of Pittsburgh that the

149

great distances between the observed quasars told us even more about the natural constants.

In an expanding universe with a finite past there are objects that are so far apart from one another that light has not had time to travel between them. This is because in its early phase the cosmos was expanding faster than light (which is not in conflict with the laws of causality!). Thus, each light wave traveling across space could only reach a part of the universe. All objects observed from a particular point lie inside what is called the *horizon* of that point. The horizon of any point is the region of space that can be causally reached from that point, that is, the space within a sphere centered on that point whose radius is increasing with the speed of light.

In cosmological theories in which the laws of nature undergo change, the values of the fundamental constants at any point in space-time are determined by all the matter within the horizon of that point. This is a consequence of Mach's Principle, according to which at least the inertia of a particle is determined by all the other matter in the cosmos. If the horizons of two points at a great distance from each other — two quasars — do not overlap but the constants within each of the two regions show the same numerical values, it follows that these values cannot be determined solely by the matter contained within the respective horizons. This, however, would contradict Mach's concept, at least in a generalized form by which *all* fundamental constants, rather than only the mass inertia, are determined by the matter inside the horizon.

Instead, the identity of natural laws in two regions with no causal connection must be traced back to a process that is part of the Big Bang itself. Tubbs and Wolfe pointed out that the horizons of the two furthest-apart quasars overlapped by only 27 percent; and therefore most of the matter inside one horizon would be unable to affect the matter in the other. Consequently, the values of the fundamental constants in the two regions would not have been able to exert any reciprocal influence. With reference to the electromagnetic fine structure constant, the strong nuclear force and the mass relationship of electron and proton, this means that these values were the same everywhere very soon after the birth of the cosmos, and that they have remained the same everywhere since.

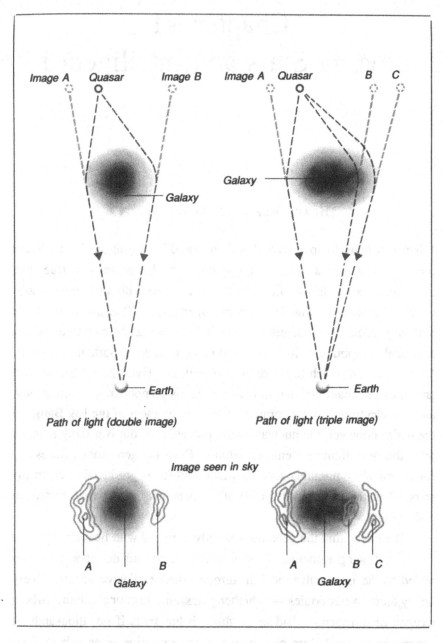

How a galaxy as a "gravitational lens" could conjure multiple images from the light of a quasar is shown by the path of the light rays (above) and the resulting distribution of light in the sky, which we would astronomically take as real (below).

Chapter VI
Exploding Stars and Intelligent Life

Heavy Elements for Biology

It is not difficult to appreciate that there could be no life on Earth without the sun. But it is a good deal less obvious, though no less true, that there could be no life on Earth or anywhere else without regular deaths among the stars — that is, without supernovas. This arises from the necessity of heavy elements for life. Take water as an example: this, a chemical compound of hydrogen and oxygen, is an important part of the foundations on which terrestrial life is built up. Hydrogen is the lightest and most abundant element in the cosmos; its components — one proton and one electron — were formed in the very moment of the Big Bang. In the main, however, the nuclear fusion processes of the Big Bang stopped after the next lightest element, helium. Even oxygen atoms, necessary for chemical compounds such as water, had to be produced later, in the stars. And the same goes for all other elements heavier than hydrogen and helium.

The gas within the galaxies was only enriched with heavier elements in the following millenia. First it was "boiled" inside stars, and then added to the interstellar medium through their explosive deaths. Every heavy atom in our bodies — whether potassium, iron or calcium, carbon, oxygen or nitrogen — had to go through hundreds if not thousands of supernova cycles before condensing in the primitive solar nebula from which, 4.7 billion years ago, the sun and the Earth were born.

Right: The schematic for the manufacturing process of cosmic matter. The death of stars–supernovas–play a central role in it.

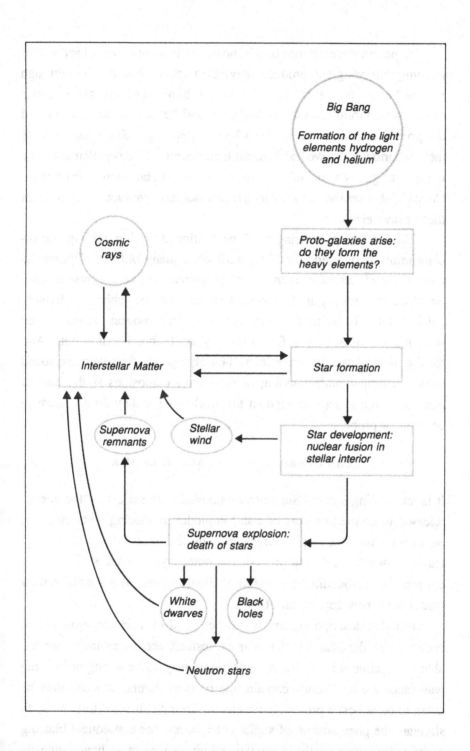

To permit the existence of life, however, it is not enough for the stars to come into being and produce heavy elements in their interiors through nuclear fusion. New stars today are not just born out of interstellar clouds of the same virgin mixture of hydrogen and helium as the first stars of the primitive galaxies over fifteen billion years ago. Since stars release their laboriously brewed and accumulated store of heavy elements, the interstellar gas is enriched with them. The next star formed from these clouds builds on the work of its predecessor and produces still more of these heavy elements.

The interstellar mincing machine continued turning, mixing and re-consuming its own end-products, until, after many billions of years the cosmic abundance of elements reached something like its present state. An effective throughput of processed matter could be achieved relatively quickly, for it is the most massive stars that die soonest. A star of ten solar masses lives only a few million years before it burns out. And for the life that was to arrive later, the only "good" stars were exploded stars. A proper understanding of supernova explosions is the aim of many research groups throughout the world — but they do not seem to be short of problems.

The Supernova Experts' Dilemma

It is an exciting story: "Supernova explosions are set off by the energy released when the iron core of a star implodes, producing a neutron star; or, in the case of a star of large mass, by the implosion of its core to leave a black hole. In the process, the outer layers of the star are flung out into the surrounding interstellar medium in order to redistribute their matter for a new generation of stars."

But this description, which can be found even in the most recent books under the heading of stellar explosions, seems, as far as we are able to examine it in an objective scientific way, to be wrong or in some way incomplete. It must contain one or more errors, and can thus be taken to be at best a provisional scenario. Despite all the efforts made to simulate the phenomenon of stellar collapse and the consequent blasting of its outer layers on the computer, researchers seem to have come up

154

against a brick wall for the moment: in numerical simulations, stars do indeed collapse as expected at the end of their evolution, but they do not explode.

"That's tough on nature," as Albert Einstein said when asked what his reaction would be if observations were to contradict his mathematical equations. Different areas of physics operate together in the description of a stellar explosion: nuclear physics, the theory of nuclear fusion, gravitational theory, particle physics, thermodynamics, the dynamics of gases, and a good deal of astrophysics.

Only in the case of stars of large mass can a supernova come about at all — our sun will never "explode." The following considerations thus apply only to stars of over ten solar masses.

There are three phases leading up to the supernova event. *Phase 1*: the "normal" life of a star — what is known as a "main sequence star," and then as a red giant. *Phase 2*: the collapse of the star in an implosion, followed by an explosion in which the larger part of the stellar envelope is catapulted into interstellar space. The after-effects are observed by astronomers in *Phase 3*— recently, for example, in one of our neighboring galaxies, and also by Tycho Brahe and Johannes Kepler in our own Milky Way: a flare-up that can last for weeks, with peaks of brightness sometimes eclipsing the light of the whole galaxy. Supernova remnants are also observed: dispersing gas clouds, as in the Crab or Vela nebulae.

The periods over which each phase may stretch differ greatly. While a high-mass star may burn for several million years as a "normal star," it eventually collapses in a split second (Phase 2) and explodes in around 10 seconds (although the glow of radiation energy can still be seen for months afterwards.)

At first a heavy star "feeds," like the sun in its present state, on its hydrogen which it burns to helium at its center at a temperature of some 30 million degrees. (This is carried out by means of the "carbon-oxygen-nitrogen cycle" first discovered by Bethe and von Weizsäcker.) The energy released in this nuclear fusion process keeps the globe of the star stable at a particular diameter: the outward pressure of radiation and hot gas balances out the inward effect of gravity.

When the hydrogen at the center is exhausted and completely converted to helium, the star first of all begins to shrink. At this point the fusion oven in its interior has cooled to the extent that it is providing less energy than is being radiated at the surface. The star can now tap fresh energy from two new sources, however: on the one hand, the process of compression sets off hydrogen burning in the outer layers of the star's gas envelope; and on the other hand, the helium at the star's center contracts to the point where, at about 200 million degrees Kelvin, the next nuclear burning stage, helium burning, is initiated, and the fusion of helium particles forms carbon via the unstable intermediate element of beryllium. As soon as the helium is used up — it provides far less fusion energy than hydrogen — the contraction process gets under way again. After helium, carbon is set burning, then oxygen and finally, at temperatures above 4 billion degrees, silicon. Silicon burning, the last stage in nuclear burning, produces the most stable of all elements: the atoms of the iron group — iron itself, nickel, cobalt, manganese, chromium and titanium. The preceding stages of burning having provided decreasing measures of energy in shorter and shorter periods of burning; iron is where the process stops. The nuclear fusion of iron atoms liberates no more energy.

At the same time the continuing contraction of the star gives rise to more layers undergoing nuclear burning in the outer regions of the star, finally and furthest in a silicon-burning layer. However, at the star's center a large core of iron and nickel has been accumulating, and this is now being further supplied with metal by the nuclear silicon source. At about one-and-a-half solar masses, its attraction reaches a critical value at which it can no longer resist its own gravitational pull. After the leisurely, almost static consumption of nuclear fuel, the star now suddenly changes pace. Phase 1 ends at the moment the star gets out of equilibrium. Now begins Phase 2. The interior of the star collapses in a fraction of a second. The remainder of the star shrinks in on itself from one moment to the next, and at ever-increasing speed. Nothing of this collapse will reveal itself to the distant observer: he will see only the later explosion. Gravitational waves and neutrinos created in the collapse may one day be recorded by special receivers yet to be made.

The collapse liberates gravitational binding energy of which about one-tenth is hurled out into space within the first second after the start of the collapse with supernova energies of 10^{51} erg corresponding to 3×10^{37} kWh. For this, however, the collapse must be halted at some point; and this comes about when the matter of the star is so tightly compressed that it becomes as dense as in an atomic nucleus. At this point an opposing pressure builds up which brings the collapse to a halt, so marking the end of Phase 2. The intermediate result is a compressed body in the interior of the star, one-and-a-half times the mass of the sun and with a diameter of 20 kilometers as dense as an atomic nucleus but with a volume of 4,000 cubic kilometers.

STELLAR EXPLOSIONS

As soon as the collapsed matter of a star comes abruptly to a halt at its center, its kinetic energy can be reflected outwards again by means of a shock wave directed away from the center and passing through all the still-collapsing outer layers of the star. Besides this, the neutrinos produced in the star's nucleus in the last hot collapse phase can now transfer energy outwards to the star's envelope.

"The result is a supernova explosion in which the greater part of the star's mass is flung out into interstellar space at a velocity of around 10,000 kilometers per second," as Wolfgang Hillebrandt of Munich's Max Planck Institute for Physics and Astrophysics recently summarized it. But Germany's supernova expert also admits: "As of today, it still has not been completely established that the processes described do indeed lead to a supernova explosion."

Theoretical physicists have been actively working on this problem since the end of the sixties, without having found a conclusive answer. Today at many scientific centers giant computer programs simulate the collapse process, duplicating in mathematical equations how hundreds of different nuclei react with each other at several billion degrees in rapidly compressing matter. Sometimes the computer does show a genuine explosion — but only under physically impossible assumptions. For realistic initial conditions and physical laws the result is: in the star's interior several layers around the iron-nickel-neutron star are again thrown outwards through "core bounce," with a velocity of approximately 30,000 kilometers per second, but only for a small distance. The shock wave itself, which rushes out from the interior at nearly 60,000 kilometers per second (20 percent the speed of light), carrying part of the core's matter with it, already ebbs in split seconds. The atomic nuclei caught up by the wave quickly dampen it and remove all its energy. The shock wave and the matter moving with it are stopped long before they reach the star's outer layer. With the computer there is still no Phase 3, no supernova.

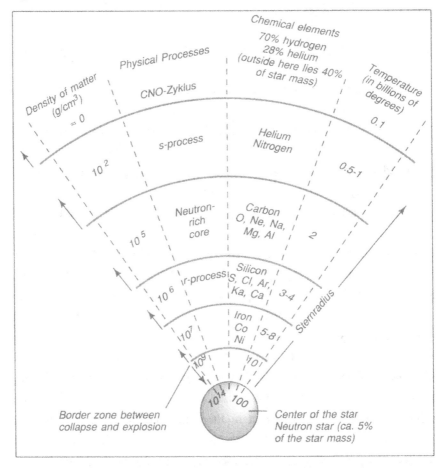

Only by the explosion of stars can the elements essential to life, which are formed in the stars' interiors, be ejected into space.

What are the physicists overlooking? What are they doing wrong? From the start it should be stressed that the "error" in theoretical understanding must lie in the time before the "bounce," therefore in the physics of the collapse itself — because what happens after that gives rise to consequences which fit amazingly well with our observations. At least three supernova-remnants are known in which neutron stars sit: the Crab Nebula, the Vela Nebula, and RCW 103. By assuming that at the heart of the complex star — on the brink of the neutron star — a shock wave of sufficient strength comes into being and travels outward as a spherically symmetric wave of explosion, then what happens can be pursued further. In less than a tenth of a second the assumed shock wave traverses the inner zone, and after several hours it has also penetrated through the outer layers of the star. Its departure velocity reaches five times the local speed of sound, a shock with Mach 5! The spherical wave races through the star raising the density and temperature at the shock front four to six-fold and igniting further nuclear fusion for tenths of a second in the various layers through which it hurries.

Finally the star's outer shell is thrown off and with it the newly "combusted" thermonuclear elements are flung into space. But we must not forget: in order for this supernova explosion to succeed, a shock wave from the interior must be assumed whose force could tear away the star's shell. At this time the energy of this supernova (approximately 10^{51} erg) still has to be fed "by hand" into the computer program. But the composition of elements in the exploding matter corresponds to that which we observe in the sun, at least with the elements that are heavier than hydrogen and helium.

"IS SUPERNOVA RESEARCH IN A BLIND ALLEY?"

Thus the error comes some place before. Is it in the collapse? In the bounce? With the neutrinos? There are many possible sources for the error and several unknowns, because here physics has advanced into areas in which natural law can no longer be proven through experimentation on Earth. Scientists anticipate there might be more energy than calculated at several points, beginning with the condition of matter under extreme compression. For the innermost collapse zone the law governing the density of atomic nuclei must be extrapolated. Then there is the rotational energy of the dying star: a slowly rotating pre-supernova star at the end of the collapse will transform itself into a nucleus which most probably rotates several hundred times per second. The distribution of matter no longer has spherical symmetry — a simplifying assumption which until now has been widely used in computer models. Yet another potential generator of energy is the magnetic field; with the collapse, the strength of the magnetic field is increased possibly up to one billion million times the Earth's magnetic field (10^{15} Gauss). Along with that the rotation also spins the magnetic field, which creates a considerable magnetic field pressure. This pressure could add to the force pushing matter outward.

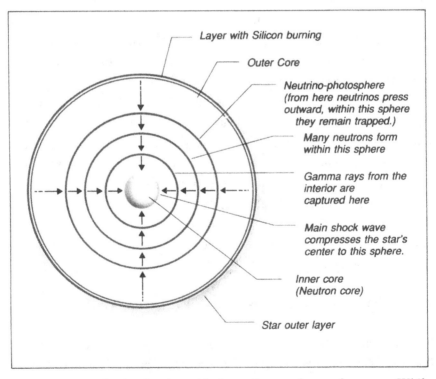

A supernova explosion begins with the collapse of a star's center. While on the outside nuclear processes (silicon) still proceed, in the interior a neutron core and energy rich emissions (gamma rays, neutrinos) form in milliseconds.

Lastly, physicists have turned their gazes to neutrino transport. For supernova theory it is decisive to know where and how the neutrinos surging out of the collapsing core deliver their energy and momentum to the star's outer envelope, and therefore to the gas which should be expelled through the detonation. To be effective, they must arrive within tenths of a second to the outer layer of the star. Neutrinos which are caught immediately in the core, inside the neutrino sphere, contribute as little to the blasting of the outer layer as neutrinos which simply pass out through the star without interaction.

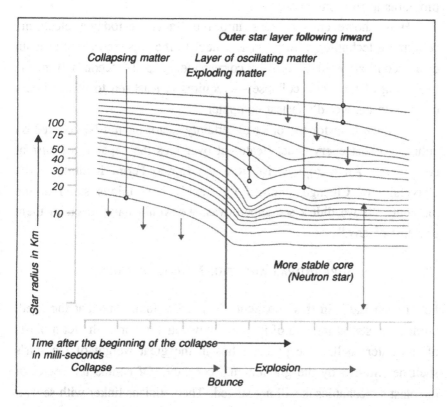

The computer simulation of a "supernova" initially shows the layers of a star collapsing. This creates a more stable core (a neutron star) against which the matter being drawn in by the collapse recoils. Instead of exploding, however, it only oscillates back and forth several times. It is still not really clear how a star goes to detonation.

Gravitational waves, which are emitted through the "bounce," could in the short term interfere with the fate of the explosion — and weaken it! The greater the collapsed core deviates from a spherical shape and becomes oscillating, the stronger pulses from gravitational waves are freed, whose energies become absorbed by the shock waves.

There is one more, entirely different possibility. Perhaps scientists have been unjustified in staring almost hypnotically at the rebounding of matter and expecting the solution to the supernova puzzle from the first tenth second after the bounce. Maybe the actual explosion occurs somewhat later in reality, and they haven't allowed the computer programs to run long enough.

Here, however, one runs into the limits of today's electronic computing technology. In order to calculate the supernova to one tenth of a second after the bounce — corresponding to 0.4 seconds from the beginning of the star's collapse — requires at least ten to fifteen hours of time on the world's fastest computer.

A quick cure for the other uncertainty factor cannot be counted on either. Almost every exact treatment oversteps the capacity of current computers. An astrophysicist from the Enrico Fermi Institute at the University of Chicago summed up frankly how things stand at the moment: "If we didn't see supernovas, we would have to claim there couldn't be any."

Supernovas and the Neutrino Fuse

For life on Earth in this connection it is most important that the aged, burned-out star at the end of its fate "blows its top" or in the least blasts off its outer shell. The problem lies in the great weight of the star's shell determined by the gravitational constant. Of course the secret of the blast's mechanics is still not solved. Theoreticians tinker with several models from which one thing becomes clear: absolutely essential for the blast are the neutrinos which are created in the star's hot interior during the collapse and release their energy and momentum to the outer shell through the detonation.

The magic key to the secrets of supernovas probably lies with the neutrino. In any case, astrophysicists are relying on its explosive force. What kind of particle is it? Neutrinos are particles of the weak nuclear force, having no or extremely little mass, able to pass through all matter. Whether a supernova explosion can come about in general depends on a delicate balance between gravity and the weak nuclear force. For the detonation to be successful, it is essential for neturinos to react quickly enough and strongly enough with the matter of the outer shell, with all of its atomic nuclei. Therefore, the momentum-rich neutrinos may not be captured in the center of the matter, because they are then absorbed immediately after their emission and are never able to reach the star's outer layers. On the other hand, they must not reach the outside too easily, as they would then simply be emitted into outer space, without having interacted with the matter in the star's shell. In neither case would the shell get blasted away.

The balance between the forces must be so precise that neutrinos can react with the shell within milliseconds and in sufficient strength for exactly the timespan in which the center of the star collapses. If the neutrinos' typical reaction time is longer, and in the decisive fraction of a second, therefore, too few neutrinos strike the atomic nuclei in the outer shell of the star, then the neutrinos leave the star without having come forcefully into contact with the star's shell. Conversely, if the reaction time is very short, and thus in the first millisecond very many neutrinos strike atomic nuclei, then virtually all neutrinos are captured in the vicinity of their birth place, the collapsing star's center. They wouldn't then even reach the outer, less strongly bound, gravitationally moderate zones of the star which are appropriate for being blasted apart.

The comparison of the time of collapse with the average peak time for neutrinos prescribes a specific relation between the weak and the gravitational forces, namely (Alpha–G) $\sim$ (Alpha–W)^4. Only when this relation is satisfied does the model for a supernova follow, allowing the elements that are the raw materials for life on Earth to pass into the universe. The above stated relation between two fundamental constants, incidentally — is it purely chance? — is identical to the one which determined the transformation of hydrogen into cosmic helium in the Big Bang (see Chapter IV).

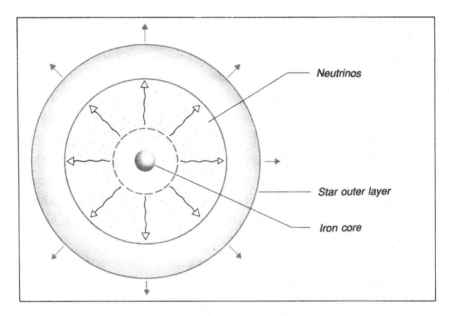

In a supernova explosion, the outer layer of a star is supposed to be blasted away by the neutrinos which form in the collapsing center.

THE SUPERNOVA OF 1054

Because supernovas appear in the Milky Way only sporadically every century, direct observations of them are especially valuable, particularly when they are connected with historical events whose circumstances we can follow-up today astronomically. Historically, just a few supernovas have been recorded: the sighting in the year 1054 is of special importance. For a long time it has been puzzling why this supernova, which was noted by Chinese court astronomers on July 4, 1054 and observed for months, was not also seen in Europe or Asia Minor. In the fifty-sixth chapter of the "Astronomical Treatise" sung-Shih it is reported: "One day in the fifth month of the first year of the reign of Chih-ho, a guest-star appeared several centimeters southeast of Zeta Tau. Over the course of a year it disappeared slowly." From this description, the "guest-star" must have

164

been brighter than all the other planets, also brighter than our morning and evening star Venus.

Today the remains of this event are on view in the Crab Nebula and constitute an object of active research. The masses of gas in the Crab Nebula are still being propelled from each other at a velocity of 1,000 kilometers per second; in its center a rotating neutron star can be observed. Explanations were long sought as to why there were no European or Middle Eastern recordings of this noteworthy event. There were plenty of guesses: months of bad weather, cloudiness, and even the limited imaginations of medieval contemporaries were credited. Finally, in 1978 Kenneth Brecher of MIT and his colleagues announced in the science journal *Nature*: "We are presenting here for the first time a source which we take to be a Near Eastern eyewitness report of the 1054 Supernova explosion."

Apparently at that time in 1054 a Christian doctor from Constantinople observed the supernova and took notes on it. The doctor, Ibn Butlan, practiced his stargazing not on account of astronomy, but out of the desire to locate heavenly causes for the starvation and epidemics which were decimating the population of Asia Minor at that time. Ibn Butlan's report, published in a collection of doctors' biographies put together in 1242, reads: "One of the well-known epidemics of our own time is that which occurred when the spectacular star appeared in Gemini in the year 446 H [April 12, 1054 through April 1, 1055 A.D.]. In the autumn of that year 14,000 people were buried in [the cemetery of] the Church of [St.] Luke, after all the [other] cemeteries in Constantinople had been filled... as this spectacular star appeared in the sign of Gemini, which is the ascendant of Egypt, it caused the epidemic to break out in Fustat [old Cairo] when the Nile was low.... Thus Ptolemy's prediction came true: 'Woe to the people of Egypt, when one of the comets appears threateningly in Gemini!'."

On account of the so-called precision of the equinoxes, a slow shifting of the Spring's points and with it of the zodiac, in the year 1000 the Crab Nebula was still in the zodiac sign of Gemini; today it is in the neighboring sign of Taurus, in the vicinity of Gemini. (Naturally, the Crab Nebula always remains in the same position in the constellation of Taurus — undisturbed by the shifting zodiac signs.) The date and position of Ibn Butlan's observation correspond exactly to the Chinese sighting from July 4, 1054. The Arab doctor's report provides nothing new astronomically, but it strengthens the credibility of astronomy in the medieval Islamic world.

DID A SUPERNOVA CAUSE THE BIRTH OF THE SOLAR SYSTEM?

In the past few years clues have grown indicating that young stars owe their creation to supernovas. With our own solar system as well, a nearby supernova exploding at a distance of a few light years seems to have been the triggering moment, causing the collapse of the protostellar fog out of which the sun and its planets developed.

If this theory holds true, then supernovas in all probability had a second function in the birth of life, aside from the production and distribution of the heavy elements. In general it is accepted that stars (and their planetary systems) emerge from the collapse of interstellar clouds of gas and dust. However, it remains unclear exactly what brings about the collapse of a cloud. An isolated cloud left on its own would most likely fly apart, rather than shrink under the force of its gravity. Only if an outside compressing force affects the cloud will it finally reach the point where its own gravity becomes strong enough to lead to collapse.

Galactic density waves have been discussed as a possible compression factor. If a compression wave moves through a galactic center, then it produces a shock wave which condenses the interstellar clouds to the size of the solar system. The creation of many stars in the spiral arms is explained in this way. In 1953, the planetologist Opik had already suggested that gas flung out explosively from a supernova could be a possible alternative to the compression of interstellar clouds.

Following the theory of supernova remnants, the shell of neutral hydrogen ejected from a supernova continues to expand at several dozen kilometers per second for ten to one hundred thousand years after the explosion. The shock wave created by the hydrogen's collisions with a cloud of interstellar material can increase its force ten-fold — enough to trigger the formation of stars.

For astronomers this process has been difficult to observe directly. US astrophysicists Nancy Morrison and David Morrison have commented "unfortunately, supernova remnants are most easily recognized when much younger than 100,000 years, while young stars are best recognized when older. Therefore, the remnant and the young stars that it spawns are rarely seen at the same time, and a pattern of association between them is difficult to establish."

Nevertheless, in 1977 W. Herbst and G. E. Assonsa of the Carnegie Institution in Washington D. C. succeeded in discovering in the constellation Canis Major a form named Canis Major R1 which fits this description. A supernova remnant, about 760,000 years old, had moved at 35 kilometers per second through an area which was filled with a series of young stars approximately the same age. Further comparable groupings were observed in the constellations Orion and Cepheus.

The conclusion that our solar system also owes its existence to a supernova "trigger" is the result of recent investigations of meteorites. Data for this explanation is provided by the composition of elements in meteors. For a while now the statistically varying distributions of several isotopes in tiny enclosures of many meteorites have been wondered about. These particular isotopes correspond to the end products of the radioactive decay of isotopes which were actually produced by nuclear fusion during the detonation phase of a supernova. How could the inclusion of supernova matter in meteorites have been achieved? Approximately five billion years ago a massive star in our vicinity could have suffered its supernova-death. Hot gas and heavy elements like nitrogen and oxygen could have met with an interstellar cloud and condensed it, so that as a result it gathered itself into a star with planets and other smaller bodies. Still during their creation, individual pieces of hot matter from the supernova could have been encased in the small meteorites in which they would have been kept unchanged until the present day.

Heavy elements and many stars owe their existence to the death of heavier stars; both are essential prerequisites for planetary life. Naturally, at least as essential is the long-term presence of a star, which, like the sun, radiates its energy uniformly for several billion years and because of this creates tolerable temperature conditions on a not too distant planet.

The sun is a so-called main sequence star, belonging therefore to the largest group of all stars. Red dwarves, blue giants, and white dwarves are the extremes of these main sequence stars into which the sun will also develop at one time. Along the "main sequence," the group of the

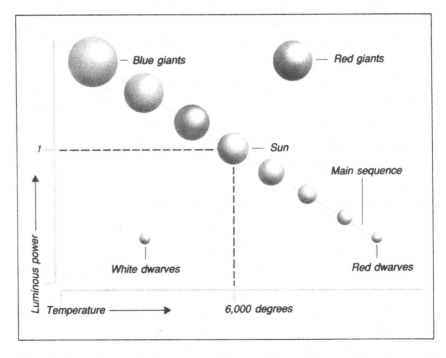

The sun occupies a middle position between the cooler red dwarves and the hotter blue giants. Since planetary systems presumably appear above all with red dwarf stars, the sun consequently belongs to the hottest stars of this kind. A cooler star than the sun could not have set in motion the energy storage system of green plants (chlorophyll).

most frequently occurring stars, our yellow sun lies in the middle, between the least dense red dwarves on one end and the most massive blue giants on the other. Because the sun has a relatively low density, it is also included among the "yellow dwarves."

So much for astronomers' colorful names. More important for our discussion are the physical differences between the red dwarves and the blue giants. As we will see, the existence of the yellow dwarf clearly depends on a specific relationship between two fundamental constants.

What differentiates blue giants from red dwarves above all is the way in which the energy they produce in their interiors is transported to the surface of the star. The energy of all stars comes from the thermonuclear fusion of lighter nuclei, predominantly from the nuclear "combustion" of hydrogen to helium. With higher temperatures the transformation of helium nuclei to carbon and oxygen and still heavier elements is also set into motion. The nuclear chain of combustion comes to an end with iron. It is the most stable chemical element in nature, with the greatest binding energy. Fusion to make even heavier elements would not release any more energy, but consume it.

The energy from the interior of the star gets to the surface through radiation (photons) or through the upward flowing of hot gas layers, while colder layers of gas simultaneously sink. Everyone who has stood before the heating element of a central heating system has felt the transport of energy through infrared radiation; the upward flux of matter — what is known as convection — is practiced daily in every cup of tea. As a third possibility, energy can be transported through heat conduction, as occurs in the warming of a metal bar.

Blue giants maintain the high temperatures at their surfaces through the energy which gets there by radiation from the interior. The chemical elements of the star gas remain fixed at a certain layer; viewed chemically the star forms an "onion skin structure." The smaller the stars are along the main sequence, and thus the cooler they are at their interiors and on their surfaces, the less radiation is produced and the less effective their energy transport by radiation. In addition, at the outer layers a "convection" layer develops in which the gas flow fluctuates, creating

169

supplemental energy for the star's surface. In this case radiation and the flow of the gas together maintain a constant surface temperature. The yellow dwarf star "sun" assumes an intermediate position between these possibilities. It has a thin surface layer with convection, which is approximately one tenth the thickness of the star's radius. The main sequence stars with the smallest mass, the cooler red dwarves, are completely mixed together with gas currents, like in a whirl; they are "fully convective." Because of this such a star can form no shells at all in its interior, and instead it is chemically homogenous.

Supernovas are as important as planetary systems for the creation of life. For the anthropic principle, therefore, the following two types of stars are of consequence:

— radiation dominated stars (blue giants) forming the onion skin structure, as they provide models for stars which are on the verge of a supernova explosion.

— (Partially) convective stars like the sun, as they stand in connection with the creation of planetary systems.

The last statement is supported by astronomical observation that red (and yellow) dwarves have far less angular momentum than blue giants. The angular momentum otherwise stored in rotation can be given up to a planetary system. This is exactly the case with the solar system: almost the entire angular momentum of the solar system, over 99 percent, is held by the planets. As Brandon Carter has recognized, the mass associated with the border between the convective red dwarves and the radiation dominated blue giants actually lies in the region of the long-lived yellow dwarves, if electromagnetic radiation forces and gravity stand in a specific relationship to one another. The electrical force controls the intensity of the radiation produced, gravity determines the total mass of the star. For this to occur, the electromagnetic fine structure constant and the gravitational constant must fulfill the relation (alpha$-$G) $\sim$(alpha$-$E)^{20}. In other words: it is only because gravity is so much weaker than the electromagnetic force that there can be in general partially convective stars with masses in the order of magnitude of the mass of the sun.

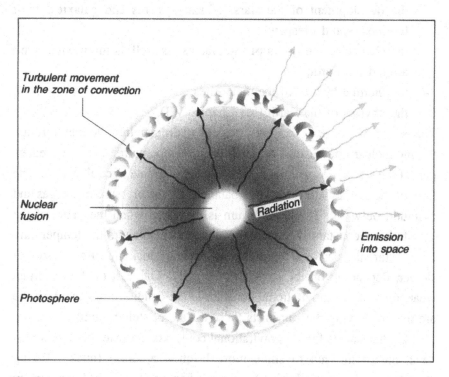

Turbulent movement
in the zone of convection

Nuclear
fusion

Radiation

Emission
into space

Photosphere

The sun is a star whose outer layer is constantly being intermixed. Supposedly, most of the stars of this kind possess planetary systems.

An alteration of the fundamental constants in another direction would remove from nature the blue giants and with them all stars capable of becoming supernovas. With a smaller gravitational constant all of the main sequence stars would be completely convective red stars. Even the larger, more massive blue giants would be cooler and red. Their gas layers would become permanently mixed together; they would be chemically homogenous and would be unable to form "onion skins" of different elements. There would then be no supernovas.

THE LUMINOSITY OF THE SUN AND THE ANTHROPIC PRINCIPLE

Astrophysical systems, which are dominated by gravity, provide the strictest tests for examining hypotheses on the variability of the gravitational constant. This becomes apparent in

— the development of the stars, of star systems and galaxies; their
luminosity and lifespan;
— the change of the orbits of the planets, as well as the moon's orbit
around the Earth;
— the thermal history of the meteorites;
— the physics of the Earth (see Chapter VII).

The speed with which a star turns its hydrogen into helium through
thermonuclear combustion depends on its temperature. For the greatest
part of the star's lifespan this process occurs in a state of equilibrium. The
flow of the star's matter inward is balanced by the pressure of the gas and
radiation outward. This equilibrium is displaced when the value of the
gravitational constant is altered. Because this also alters the temperature
of the star, the star will consume its nuclear stockpile either faster or
slower, depending on the strength of gravity. One test of the historical
weakening of the gravitational constant is the condition under which we
are now observing the sun in its present state of stellar aging.

Higher values for the gravitational constants in earlier periods would
have caused the sun to shine more brightly at those times. Taking
Dirac's hypothesis as a basis (that the weakening of alpha–G is inversely
proportional to cosmological time), then the sun's luminosity reacts,
as Edward Teller had already noted in 1948, to the seventh power of
changes in alpha–G. This is a drastic effect that would not be without
consequences for the climate of the Earth. It seems obvious that the
history of the Earth's climate could be taken as a test of the weakening
of the gravitational constant. Unfortunately, despite several computer
models, there is still no truly dependable model for the climate which
would be accurate over billions of years. Because so much is unknown
in biology, meteorology, and geophysics, this test is still dependent on
too many assumptions.

Happily, we know the sun in many aspects essentially better than
our own Earth. So the effects of a change over time of gravity on the
fate of the sun can be determined more easily. In 1964 two American
astrophysicists, P. Pochoda and Martin Schwarzbild (and George Gamow
in 1967) conducted a study of the development of the sun from this

viewpoint. Their results were noteworthy. If Dirac's hypothesis holds true, then the universe must be at least thirteen billion years old. Otherwise the sun could not have existed for 4.5 billion years or have arrived at its present luminosity. In 1964 the estimate of the age of the universe was much smaller than it is today. The current assumption of about twenty billion years for the age of the cosmos is not in especially great conflict with Dirac's hypothesis. A repetition of the calculations would be in order.

A direct measurement of changes in the sun's luminosity over time has also been attempted. The first measurement of the sun's total energy was conducted in 1837 by Claude Pouillet, a Frenchman. He coined the term "solar constant" for the energy quantity of 1.79 calories, which is the amount of energy that falls per second from the sun to every square centimeter of the Earth. With the invention of the bolometer — a kind of thermometer — by Samuel Langley of the Smithsonian Astrophysical Observatory in Washington in 1881, the measurement was refined. Finally, Charles Greely Abbot, Langley's successor, conducted the first systematic, daily measurements of the solar constant at several places around the world. They showed that between 1923 and 1950 any long-term deviation amounted to at the most 0.17 percent. More recent evidence from the years 1975 to 1979 gathered by Bill Livingston and colleagues at Kitt Peak Observatory allows for a long-term diminution of the solar constants of a maximum 0.6 percent. It is not possible to bring these results into agreement with a change in the gravitational constant.

Alterations in alpha–G would also have to influence pulsars. The rotating neutron stars, which in a space of milliseconds emit radio waves like some kind of lighthouse, would have to slow down their rotational velocity. This would then increase the number of slowly rotating pulsars. An evaluation by the astronomer V. N. Mansfield yielded a value for gravity's reduction over time, which at the present does not stand in contradiction to Thomas van Flandern's measurements (see Chapter IV).

Meteorites are also full of evidence. One can read the history of the early solar system in their internal structure. If the sun shone more strongly on them at some time they would have to have been warmer then. This would have brought about a more rapid gasification of certain isotopes and with it varying compositions of gaseous and solid isotope mixtures.

In the next chapter we move to a discussion of the geophysical effects of a gravitational constant which varies over time.

Chapter VII
Earth and Cosmos: Geophysics and the Anthropic Principle

THE EARTH AND VARIABLE GRAVITY

If the form and structure of a system are principally determined by gravity, which is true for all large bodies, and if the gravitational constant varied, this form would also change. The consequences would be especially important for the Earth. Perhaps they already occurred; the real question is whether we can trace the countless transformations of the Earth and especially the Earth's surface unequivocally back to a variable gravity whose strength has diminished over time.

The distances of the planets from the sun would be directly affected. The stronger the gravity, the more closely the planets would move in their orbits around the sun. The same would hold true for the moon and the Earth. According to Dirac's hypothesis the distance from the Earth to the sun would be increasing by about 10 meters per year. Therefore the Earth would be distancing itself slowly from the sun in a very tight spiral. For the moon the spiral would turn out to be proportionally smaller: its distance from the Earth would increase by only 2 centimeters every year. Measurements of the distance between the Earth and the moon made with laser beams, which are projected at the moon and reflected back by a mirror set up there during the Apollo flights, have an inaccuracy of nearly half a meter. This contribution to "moon flight" is not directly measurable. Unfortunately the chances for an improvement in the future are not great. The high and low tides and their corresponding internal

175

friction already cause the moon to move a greater distance away from the Earth every year. Because of their complexity tidal effects defy exact observation and calculation. Thus, so small an effect as two centimeters per year can scarcely be filtered out dependably. Shapiro's radar measurements of the distances between the other planets and the sun are not any better. They yield inaccuracies of approximately one kilometer. The reason for this lies above all with these planets' unfamiliar topographies.

But back to the Earth and the moon. The permanent movement of low and high tides moderates the Earth's rotation through friction. Because of this the Earth turns increasingly slowly. The constant reduction of Earth's turning was discovered at the beginning of the 18th century by the English astronomer Edmund Halley. He was the first to realize that a comparison of the data on solar eclipses noted over 2,500 years ago by Babylonians and Egyptians shows a systematic chronological mistake. Halley could correct the mistake by assuming that the Earth rotated faster in earlier times. In principle the fossils of certain snails and coral allowed for a second comparison. Their growth is dependent on the rhythm of day and night. Their calcium shells show daily rings similar to the yearly rings with occur in trees. As later investigations indicated, there are still problems of interpretation: "Although the growth rings in fossils promised to become a worthwhile source of information, recent difficulties in interpretation arose, such that quantitative conclusions can not yet be drawn" (Canuto). A third indication is provided by comparisons between the moon's movement and "atomic time," a measure of time, which is not determined by gravity, but rather by electron oscillations in the atomic shells of selected atoms (Krypton, for example).

Taken together these facts indicate that the Earth day is prolonged by two milliseconds every century. Every attempt to explain these changes is influenced in any case by theoretical models and assumptions, so they do not contradict the hypothesis of a cosmological reduction of the force of gravity.

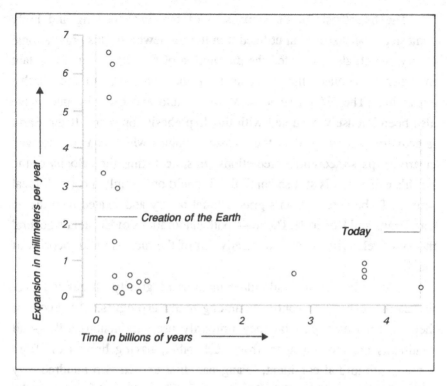

After measurement it cannot be ruled out that the Earth expands a fraction of a millimeter every year. Whether this is brought about by a diminishing gravitational constant, however, is questionable.

THE EXPANDING EARTH

A weakening of the gravitational constant over time would show itself more clearly in an expansion of the Earth. With the passing of time, the matter on the Earth would be more weakly held together — the Earth would expand and its radius would grow. Predictions for the rate of expansion based on Dirac's hypothesis and similar models fluctuate between hundredths of a millimeter and three tenths of a millimeter per year. A series of geophysical measurements suggest that the Earth grows by a half millimeter yearly. This number is supposed to represent a moderate rate of growth for the entire history of the Earth.

177

The hypothesis of the expansion of the Earth is long and hotly contested. Pascual Jordan utilized it in the framework of his gravitational theory, which also predicted the diminution of the gravitational constant in order to connect the movement of the continents to the Earth's expansion. The NASA scientist V. M. Canuto and his colleagues have also been intensely occupied with this hypothesis for years. If the Earth is actually expanding, then the measured values which can be observed clearly surpass theoretical predictions. In spite of this, the velocity of the Earth's expansion is still so small that it could only imply a cosmological source. To be sure, Jordan's gravitational theory and related theories by Carl Brans and Robert H. Dicke — both elaborations of Einstein's general theory of relativity — are practically out of the race as valid theories of gravity.

Independent of the individual theories of gravity, discussion over the Earth's expansion continues among many geologists. As proof for their theories of expansion they primarily use continental drift — all continents are moving apart from each other, having broken off from a single primordial continent, "Pangaea," that existed two hundred fifty million years ago. To explain the movement of continental plates over the Earth's softer mantle by the Earth's expansion, one must inevitably assert that the Earth has grown over 20 percent larger in the past two hundred million years. That is five centimeters a year on average. The large primordial continent then would have covered the entire young Earth at the time.

Almost everything we know about the Earth and the planets argues against such ideas. At one time in the primordial continent Pangaea there was an "opening," a primordial ocean called "Tethys," whose existence is denied by the expansionists. In comparison, other planets in the solar system — the Moon, Mars, Mercury — have not expanded at all in the history of the solar system. Finally, no convincing mechanism for the Earth's rapid swelling has ever been given. A variable gravitational constant could at best account for a fraction of a millimeter per year. The rest is atrributed to the creation of new matter by the high temperatures and pressures in the Earth's interior — this only happens on Earth, however. These speculations are rightfully discounted as "geo-mysticism."

Altogether one can say that continental shifting caused by movements in the Earth's mantle (convection) fits better with observations. A small expansion of the Earth would be completely submerged by the effects triggered by convection in the mantle. On tectonically dead planets such as Mercury or the Moon, the planet's mantle itself does not move. There expansion would stand out more clearly — but it has not been observed.

THE COOLING EARTH

The temperature of the Earth's surface would also vary if the gravitational constant were a function of time. Despite the above-mentioned problem with calculating an exact climatal course, several plausible trends can at least be determined. A stronger gravity in earlier cosmological epochs would have brought the Earth somewhat closer to the sun, and accordingly the sun would have certainly radiated more energy toward the Earth. Taken together the two would have implied a higher median temperature of the Earth's surface in the past.

The difficulty of making an exact valuation of the Earth's temperature in the past comes into question here: how much of the solar energy emitted is immediately reflected directly back into outer space by the atmosphere? The "reflection gradient" of a surface is termed its "albedo." A uniformly closed, extremely bright cloud cover would give the Earth a very high albedo, between 90 and 100 percent. On cloudless days over dark land masses the albedo is very tiny. In 1948 when Edward Teller investigated the influence of gravitation on the surface temperature of the Earth for the first time, he assumed for the sake of simplicity a constant albedo for Earth's entire history.

His second assumption concerned the age of the universe: the younger the universe, the more severe the effect of a change in gravity would be on the Earth's heating. In 1948 this age was taken to be 1.9 billion years. Starting with today's average Earth temperature of 63 degrees Fahrenheit one can calculate with the help of the Dirac hypothesis

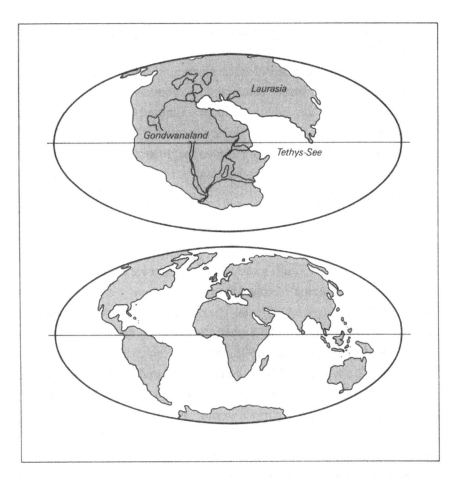

As "proof" for the expansion of the Earth, many geologists point to the distribution of the continents 250 million years ago. At that time, the land masses were still joined together in the supercontinent Pangaea. If gravity has diminished over time, this could have contributed to the Earth's expansion and to continental drift along with it.

that 2.1 million years ago the Earth's surface would have been boiling — this is in pretty clear contradiction with the planet's history. Two million years ago there were amphibians and plant life over all the continents, which naturally would have been impossible at boiling temperatures.

More than thirty years after Teller's calculations, the real age of the cosmos is now considered to be twenty billion years. In spite of this, his argument remains in force, if a constant albedo for the Earth is still assumed. The temperature must still have been higher in the past because of a changing gravitational constant but the change would have been somewhat slower: in the Cambrian period the temperature would have been 100 degrees and the critical boiling point would have been crossed 2.2 billion years ago.

This value still stands in contradiction with temperatures computed by geochemical analysis of flint stone from the pre-Cambrian period, three and a half billion years ago; at that time, as indicated by the geological analysis, it was not hotter than 140 to 200 degrees. Also in the pre-Cambrian there were already primitive life forms adjusted to such temperatures. The oldest of these life forms, primarily blue-green algae, have been found in fossils in the South African Onverwacht. In the face of this contradictory evidence, Dirac's hypothesis can only be saved, as astronomers W. Eichendorf and M. Reinhardt from Bochum conclude, if one assumes that the atmosphere would have protected the Earth from overheating. With a very great albedo, the ancient atmosphere would have acted like a refrigerator and in that way must have prevented the deadly heating up of the Earth's surface. However, this is unrealistic. Several computer simulations by the American physicist Michael H. Hart and other data indicate that the primordial atmosphere contained higher portions of carbon dioxide, methane and ammonia than today. These gases would have brought about a "greenhouse effect" which would have further raised the surface temperature instead of lowering it. Therefore, Dirac's hypothesis of a gravitational constant that changes over time must be viewed as highly improbable above all because of the results of the geo-chemistry of the primordial Earth.

Attempts to get a more precise mathematical handle on the Earth's cooling have not yet progressed far enough to enable the exploration of the possible influences of diminishing gravity. Today, the Earth gives off heat with an average output of 40 million megawatts, equivalent to the output of 20,000 atomic power plants. At that level the Earth's radiant energy, as geophysicists Henry Pollack and D. Sprague determined, lies slightly below the average value for the last 180 million years of 43 million megawatts. Indeed, at all times the total emission over the entire space of time fluctuated between 39 and 52 million megawatts.

Although part of this energy comes to the surface from the several thousand degree interior of the Earth, it appears that an even greater portion comes from the thermal power of the atomic decay processes within the crust of the Earth. Radioactive elements, principally isotopes of uranium, thorium, and potassium, are deposited abundantly in the Earth's crust. They are all nuclear "slow burners," because they have half lives of between 0.7 and 14 billion years. Mathematical models for the Earth's cooling nevertheless allow for a few conclusions about the chemistry of the primordial Earth. They indicate that a part of the thermal energy still being released today stems from the early phase of the creation of the planet. It is unclear how great a portion this is. The following conclusions are certain:

— The Earth cools itself very slowly, every ten million years a maximum of two degrees. This makes the Earth, as Henry Pollack puts it, the "best energy storage facility imaginable."

— With cooling, the Earth's mantle hardens and currents in the Earth's interior slow down.

It must be stressed that neither the expanding nor the cooling of the Earth indicate a slow diminishing of the force of gravity. As shown earlier, geochemical data about the young Earth's surface swung the pendulum against this idea. That this proof to the contrary is not perfect lies with the complexity of the Earth as a system. To abandon so fundamental a physical law as Dirac's on account of it would be foolhardy. Unaltered by this is the fact that a changing gravitational constant, now as before, could influence essential processes in and on the Earth: thermal radiation, expansion, the magnetic field, climate and the creation of life.

RADIOACTIVE ELEMENTS AND THE STRONG INTERACTION

Radioactive elements have heated the Earth's surface for billions of years. Compared with the sun's radiation, the Earth provides only an infinitesimal contribution to the warming of its surface. However if they don't keep the cradle of life warm for us directly, radioactive elements nevertheless are coupled in an interesting way with the existence of life.

Atomic nuclei of a specific size split by themselves. Chemical elements become radioactive when their atomic nuclei transform into other nuclei, especially when they disintegrate. The transformation process engenders elements with lighter nuclei and radiation. Henri Becquerel discovered the phenomenon of radioactive emission in the element uranium in 1896. Only the heaviest elements disintegrate by themselves. In nature this affects all nuclei with more than 93 protons. With uranium (with 91 protons in its nucleus) and a series of lighter elements, whether they decay and become radioactive depends on the neutrons in the nucleus as well as the protons. With more than about 240 nucleons in the nucleus, as with uranium-235 more or less, every nuclear configuration breaks apart.

Two forces control the bonding and size of atomic nuclei: the strong and the electromagnetic interactions. They both work in opposition to each other; the equally charged protons repel each other electrically, while the strong nuclear force overcomes this repulsion — at any case over extremely small distances — in order to bind protons and neutrons together. Even in nuclear physics are there "coincidences" which are nonetheless indispensable for the existence of life and therefore are of anthropic importance. These "coincidences" concern the size proportions of several particle masses and their relationship to electric and strong forces: the most important phenomena of nuclear physics, as Brandon Carter noted in 1970, depend on the following four "coincidences" — coincidences of nuclear physics:

$$(\text{alpha}-\text{S})^2 \approx 2 \, \frac{(\text{neutron mass})}{(\text{pion mass})}$$

$$\frac{(\text{neutron mass}) - (\text{proton mass})}{(\text{electron mass})} \approx 2,$$

$$(\text{alpha}-\text{E}) \approx \frac{(\text{neutron mass})-(\text{proton mass})}{(\text{pion mass})}$$

$$(\text{alpha}-\text{S})^2 \approx 1/(9 \, \text{alpha}-\text{E}).$$

With the help of the tables in the appendix these "coincidences" can be easily calculated. These relationships are not "forced" by any physical theories and they are certainly necessary preconditions for the existence of all chemical elements that life requires. Thus, as our life form requires heavy elements in order to exist, so are we dependent on the laws of microphysics which "make" these elements. Important qualities of atomic nuclei are dependent on these four coincidences of nuclear physics, in particular on the strength of the strong interaction (alpha–S). Although the strong interaction is appropriately named and in action is the strongest of the four interactions of nature, its power only suffices strictly in order to overcome the electrical repulsion between protons and to hold the particles of atomic nuclei together.

If the strong interaction were somewhat weaker, that is, if alpha–S were somewhat smaller, it could not even hold two protons in a simple helium nucleus together. For such a reason, the heavier elements could not exist. There would be only a single element in the cosmos, the simplest element with one proton in its nucleus — hydrogen. Without the heavier elements, however, there would be no life.

And how would the world look if a stronger alpha–S held the world together internally? There would certainly be much larger atomic nuclei than uranium and trans-uranic elements: perhaps almost arbitrarily large. With that there would be no radioactivity, at least as we know it. If alpha–S were just a few percent greater, then it would have had a decisive influence on the production of helium in the Big Bang: all hydrogen would have been changed directly into helium from the beginning. In that case there would never have been any water in the universe.

Now we return to Earth — the only known carrier of an intelligent species. Whether and how its existence depends on special evolutionary conditions, especially on cosmic catastrophes, shall be examined in what follows.

CELESTIAL CHAOS AND TERRESTRIAL CATASTROPHES

The development of life was advanced in all probability by catastrophes — sometime during the ice ages. Were the ice ages started by interstellar clouds? Did the dinosaurs die because a giant comet struck the Earth?

Was biological evolution influenced by violent short-lived astronomical events? In so far as being essential for the evolution of intelligence, such events must also have occurred in the interplay of the natural laws.

We have already described how much our actual terrestrial existence is also dependent on the "good behavior" of the cosmos. The cosmic background radiation must have been sufficiently cool, and must still be so today; the sun had to be the most peaceful manner of star possible, which the main sequence stars are. Otherwise the Earth would have been afflicted again and again by celestial catastrophes, which would have interfered with the terrestrial climate and thereby radically altered conditions for terrestrial life.

The influences of unusual astronomical events on the development of terrestrial life are discussed over and over with many speculations. Aside from the fantasies of the end of the world, which creep up for instance around comets (such as Halley's comet in 1910), there is proof that a series of celestial catastrophes altered the course of terrestrial events. Just how far they were also the triggers for evolutionary developmental thrusts is not yet analyzed. Therefore let us suppose that to some extent the ice ages pressed the civilizing development of Man in a certain direction; such a pressure would not have been set into motion in a more tropical climate. The passage to a technological civilization was possibly furthered by it.

Catastrophes were already viewed as determinate of evolution even before Darwin started his research. Georges Cuvier believed that evolution was originated by a series of catastrophes *alone*, through which in each case many species would disappear while other species would be advanced. Darwin's evolutionary theory of natural selection — the survival of the fittest at the time — overtook Cuvier's ideas. It must be noted that, in part, Cuvier's ideas have come into favor again — a kind of "neo-catastrophism," if you will — without contradicting Darwin. The dinosaurs are also partly to blame for it. Today it is indisputable that several animal species, such as the dinosaurs, were driven from the world within a relatively short period of time, probably through some kind of catastrophe. It is not out of the question that decisive evolutionary processes were first set into motion by catastrophes. This doesn't stand in contradiction of the usually dominant factors of mutation and selection. When a series of creatures suddenly dies out, the other surviving species can expand into the newly freed up "living space." If terrestrial catastrophes, whatever their cause, bring about the further development of life or even make them possible at first, then they exemplify that two things must interact in the universe: first, the natural laws with all their anthropic "coincidences," and second an entirely special course of evolution.

In the last two million years, there were several, at least six, mass extinctions of biological species. The dinosaurs are the first and best known example. What gave rise to their obliteration is still disputed. Clearly, there are many candidates:

— terrestrial causes: changing climate and temperature fluctuations, increased volcanic activity, flooding of coastal shorelines; changes in eating habits let the calcium shells of eggs become stronger and firmer — the young could no longer break out of their hard egg shell prisons;

— extra-terrestrial causes: a comet, a meteor or asteroid, an eruption of solar activity, or the passing of the sun through an interstellar cloud; a climate crash initiated by a nearby supernova.

What really happened 65 million years ago in the transition from the Cretaceous to the Tertiary period is only arrived at indirectly. What is certain is that the dinosaurs disappeared with lightning speed from the surface of the Earth, geo-historically speaking, although until that time, they were, with multitudes of species, the most important herbivores of a class of animals which can still be found in southeast Asia today. Paleomagnetic measurements of the thin geological layer with dinosaur fossils suggest that within one thousand years they had all died out. The misunderstanding should also be cleared up that dinosaurs died out because they were badly suited to their environment, being in a hypertropic "evolutionary dead end." Many dinosaurs had brains comparable to those of existing animals and birds. The small ostrich-like dinosaur Dromiceiominus "was probably the most intelligent creature on

186

our planet at the time," writes Wallace Tucker of the Harvard Smithsonian Center. This animal species grew to be one hundred pounds, had a large brain, and was actually a faster runner than the ostrich of today, "hardly a candidate for extinction." And yet this species also disappeared along with two thirds of every other kind of animal existing at that time.

Dramatic, short-term changes in the Earth's climate — the plausible starting point of global species deaths — could easily have terrestrial as well as extra-terrestrial causes. Complicated interactive processes between land, water, and ice surfaces are also causes of climatal fluctuations. Through increased volcanic activity a great quantity of ash and dust could be dispersed into the upper atmosphere; because of it the sun's rays would be held back. Over longer periods the climate is naturally dependent on the movement of the continents, the volume of the oceans, the glacial regions and the size of the land masses. Also, the periodic variations in the Earth's normally almost circular orbit probably brought about the first ice ages in the early history of Earth. Yet, from all these influences the climate changes only very slowly. Volcanic eruptions during the last two million years caused no mass extinctions similar to those of the dinosaurs. Purely terrestrial causes of the disappearance of the dinosaurs therefore seems rather improbable. Short-term astronomic influences, similar to catastrophes, could have worked together with biochemical factors on Earth to the disadvantage of the dinosaurs.

COSMIC RAYS AND BIOLOGICAL EVOLUTION

There are several indications of extra-terrestrial influences of catastrophic proportions. In 1980, the physicist Luis Alvarez and colleagues found that exactly at the geological layer coinciding with the extinction of the dinosaurs, iridium, a rare element scarcely appearing in the Earth's crust, is concentrated. Iridium appears abundantly in extraterrestrial matter (meteorites), approximately 1,000 times as frequently as on Earth. Their observation implies that 65 million years ago a great quantity of extra-terrestrial matter crashed on Earth within a short amount of time. Alvarez presumes that the source of this iridium accumulation is the collision of an asteroid of perhaps 10 kilometers in diameter into the Earth.

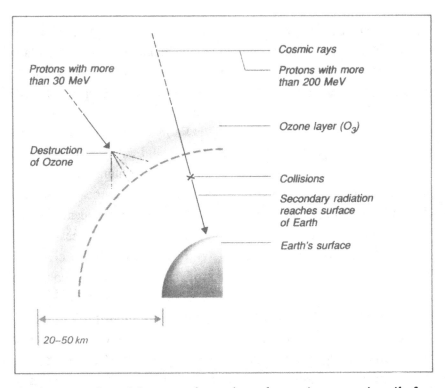

The disintegration of the ozone layer through cosmic rays, primarily fast protons and gamma rays, as well as energy rich photons. Protons with more than 30 million electron volts (MeV) reach the ozone belt; with an energy of over 200 MeV their secondary radiation penetrates to the Earth's surface. Ozone is destroyed and nitrogen oxide is produced in the process.

Right: Life on Earth and its evolution is intricately connected to terrestrial as well as extraterrestrial influences. Cosmic catastrophes could also have contributed decisively to the creation of intelligent life forms on the Earth.

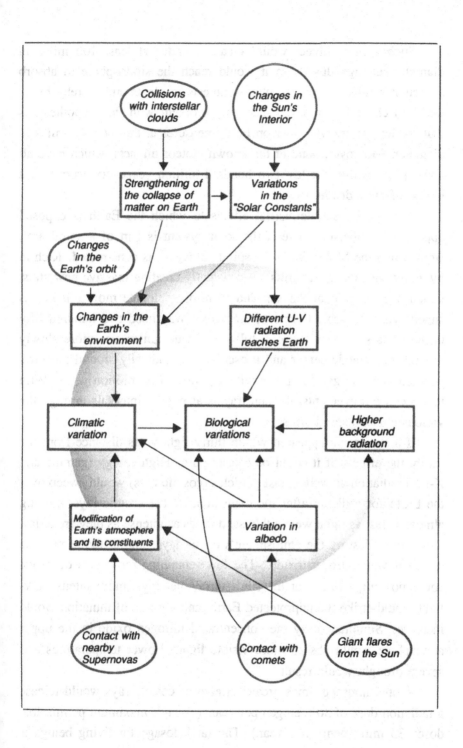

Such a huge object would create a crater at least 100 miles in diameter; enough dust from it would reach the stratosphere to absorb sufficient sunlight to reduce photosynthesis in plants and thereby break the food chain for animals. The only problem with this hypothesis is that the largest known crater on Earth measures at the most 60 miles in diameter. Moreover, none of the known meteor impacts, which have all taken place within the last two hundred million years, correspond to a period of mass deaths.

The special extraterrestrial effects to which the Earth is exposed appear more plausible here if the solar system is part of a spiral arm, from which the Milky Way — a spiral galaxy — is constructed. Rich in hydrogen gas, the spiral arms are constantly creating massive new stars, which age faster than the sun and therefore detonate more quickly as supernovas. The spiral arms of the Milky Way need two hundred fifty million years to complete one full turn. Our sun moves more slowly around the galactic center and it oscillates periodically around a central position near the galactic plane with a period of 50 million years. With its eccentric movements, the sun travels at regular intervals through the various galactic spiral arms.

If a supernova appeared within thirty light-years distance from the Earth, the effects of it could be catastrophic. High energy (gamma and X-ray) radiation as well as fast particles (cosmic rays) would sweep over the Earth for millenia after the explosion. If the cosmic rays, moving almost as fast as light, were a thousand times as intense as they are today, they would destroy the atmosphere's ozone layer and at the same time enrich it with nitrogen oxide. The few remaining molecules of ozone could no longer filter out the sun's ultraviolet rays, more intense UV-light would strike the unprotected Earth, and the rate of mutation would increase. Simultaneously the concentrated nitrogen oxide in the upper atmosphere would absorb more visible light. Lower temperatures and severe drought would result.

A one thousand times greater stream of cosmic rays would release a radiation dose of 30 roentgen per year. (Today's maximum permissible dose: 35 milli-roentgen a year.) The fatal dosage for living beings is between 200 and 700 roentgen. Indeed, a dose accumulated over decades could very easily be fatal, especially if a deterioration of the general environmental conditions took place at the same time.

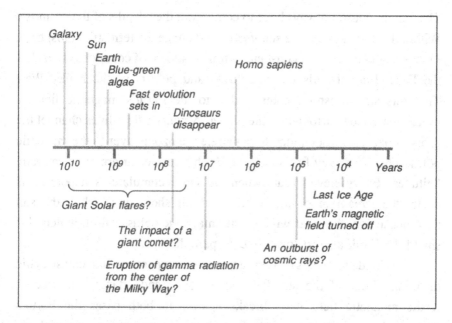

Have heavenly catastrophes influenced evolution on the Earth? This diagram shows several incidents in Earth's history.

How often can the sun come closer than thirty light-years to a supernova? The astronomers answer: every fifty to one hundred million years the sun runs into one of the spiral arms and travels through it for about ten million years. There is a 50-50 chance that the sun will then come into the threatening vicinity of a supernova. In reality, the sun (and with it the Earth) has often been affected by such an occurrence in its history, beginning at its "birth," when a nearby supernova forced a primordial cloud so strongly together that it collapsed and allowed the sun and its planets to be formed.

An explosion, which would even overshadow the effect of supernova explosions, could have taken place 140 million yeras ago in the center of the Milky Way. Some astronomers accept that at that time a massive dose of gamma rays struck the Earth and destroyed the atmosphere's ozone layer.

COLLAPSE OF THE EARTH'S MAGNETIC FIELD

The danger doesn't always have to come from the "depths of the cosmos." Within the solar system the sun itself can damage the terrestrial biosphere. In "normal operation," a constant, harmless stream of cosmic rays reaches the Earth from the distant supernovas and pulsars of the Milky Way. That this stream isn't greater is due to the Earth's magnetic field, a protective screen surrounding the planet. Magnetic field lines channel the charged particles fast enough to penetrate the ozone layer to the magnetic poles. Therefore they first reach the Earth's surface at arctic or antarctic latitudes. Every eleven years when the sun accumulates sunspots in its cycle, the stream of cosmic particles diminishes, since when the sun becomes active, the solar wind pushes magnetic fields with it, which also shield the Earth against lower energy particles.

Two kinds of events can bring this normal operation into catastrophic disorder. First, if the sun falls out of sync and "flares," or second, if the magnetic field itself breaks apart. In both cases the Earth's protective magnetic screen fails — with consequences for terrestrial life and evolution. With solar flares, which usually appear simultaneously with increased sunspot activity, the Earth is inundated for a short time with many low energy protons. As a rule they are not fast enough to penetrate the ozone layer and seriously damage it. Only occasionally is a solar flare energetic enough to strike through to the Earth's surface; this happens about seven times every solar cycle (22 years). It is not certain whether the sun with such flares is really capable of causing a mass extermination in the style of the dinosaurs or other evolutionary catastrophes on Earth. It is known that other stars, especially the so-called M and K stars, eject flares one hundred times the size typical of the sun.

Certainly, the more moderate solar flares can also do considerable damage if *at the same time* the Earth's magnetic field is shut off.

Relatively frequently, on the average of every two hundred thousand years, the magnetic field of the Earth falls into great disorder. In the last 2.5 million years this has happened ten times. The shutting down process really represents a transitional phase by which the magnetic field reverses direction and during which the magnetic field strength sinks to a much lower value. At such a time lower energy protons can reach the Earth, in which case protons from large solar flares can endanger the unprotected planet. "It is very interesting to note that recent Japanese work on field reversals from studies of deep-sea sediments show that the field was virtually switched off for 10,000 to 20,000 years just over 1 million years ago. The fact that evolution threw up Man at about this time makes for an interesting coincidence" (Arnold Wolfendale).

If in one of these phases protons, like those in cosmic radiation, came to the Earth from a solar flare with great energy, they would produce through their interactions in the atmosphere "secondary radiation" of 1,000 roentgen. "It is very likely," Wolfendale continues, "that a dose of this magnitude at any stage in the past 1,000 million years would have had a significant effect on the evolutionary track." The inhabitants of the Earth are the results of evolution and naturally of all cosmic effects, especially those which brought about terrestrial catastrophes. Perhaps these catastrophes were "indispensable," because they provided jolts which were decisive for evolution, in which case they must have been well dispersed in order for life not to be simply wiped out completely. This "almost" draws very precise boundaries for the permissible history of evolution. The scaffolding of nature with its anthropic cross-connections is only a part of the foundation of our existence. Another obviously necessary role is played by an evolutionary time-table — a subtle combination of quasi-statistical evolution and catastrophe — which brought about *homo sapiens* as observer of the universe.

THE NATURAL WORLD IN POWERS OF $N=10^{80}$

The construction of the physical world is made up of notably different levels, from atoms to cells, organisms, planets, stars, galaxies and finally to the universe. Only the absolutely smallest components, quarks and leptons, appear to contain no further substructure (of course some physicists meanwhile are already seriously contemplating "subquarks"!). Only the largest structure, the universe, allows for no further, still larger unity. Every single range of components, every level, is described by a different physical theory, and "so it is not always appreciated how intimately they are related," state Bernie Carr and Martin Rees. Indeed, many aspects of the natural world which were essential for the development of life are dependent in a fragile way on specific "coincidences" among the fundamental natural constants. On the other hand, it is also noteworthy how many fundamental qualities of the world are determined by gravity and a few micro-physical constants — not by accident, but rather as a consequence of simple physical arguments. Because it is the intention of this book to recognize "accidents" in the construction of the physical world which are preconditions for the existence of every life form, they in turn must be separated from those qualities which are not accidental. Many relations constitute physical necessities independent of whether or not life exists.

The size and weight of an object are typical qualities which are determined by physical laws. It is the "scales" of natural objects, measured by these standards, which are the measure of nature: their size, measured against the atomic radius; their mass, expressed in atomic mass (the mass of protons). An example is the size of terrestrial life forms. How large can a creature be which lives on the surface of a planet? The size of its body represents a compromise between the stability of its bodily structure and the effect of gravity on it. Both phenomena are related by specific natural constants. On one side, the physics for the binding of biological tissues is expressed by the electromagnetic fine structure constant. On the other side, the typical size for a body is set by the surface gravity of a planet. If one rationally expects that a planetary life form should not fall to pieces if it falls down, then from all this a maximum size can be determined for a person to have in order to wander more or less assuredly over the Earth. This is only *one* argument for how the scales of nature are determined. It would lead us too far afield to get engrossed in another one. It suffices to hold as a fact that all the scales can be characterized by the constants for gravity (alpha–G) and the electromagnetic force. They are summarized in the diagram on page 195. [Alpha–G is replaced by the number $N=10^{80}$ there, a simple conversion with the help of the relation $N=1/(\text{alpha}-G)^2$].

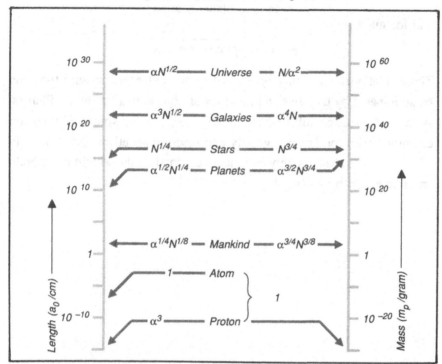

In nature, Mankind occupies an average position, lying in the "golden mean" between microcosm and macrocosm. This refers however only to external qualities: size (left scale) and weight (right scale). The diameter and mass of an atom, a_0 and m_p, serve as the standard units. It is noteworthy that all the dimensions can be expressed by electrical (alpha-E) and gravitational (alpha-G) forces. [For the number of particles in the universe, N, take $N=1/(\text{alpha-G})^2$.] (Table from Carr & Rees, 1979).

Several "amusing cross-connections" (Carr and Rees) can be derived from the diagram of the most important sizes and masses. Thus the size and mass of Man can be expressed as the "geometrical mean" (the root of the product) of two other characteristic sizes and masses. The size of a planet lies in the middle between an atom and the universe; Man occupies a middle position both in size and weight between atom and planet. The same relation holds for length

$$\text{planet} \sim \sqrt{\text{universe} \times \text{atom}},$$

and for mass

$$\text{Man} \sim \sqrt{\text{planet} \times \text{atom}},$$

$$\text{Man} \sim \sqrt{\text{planet} \times \text{proton}}.$$

These relationships cannot be viewed as coincidences, because they can be anticipated by the laws and qualities of objects familiar to us. Planets, stars, and galaxies fulfill these anticipations, so to speak, if they are examined with the laws of gravity and electromagnetism. As shown, in this connection, Man is only an object burdened by the weight of its body and threatened by its fragility.

Chapter VIII
The Uniqueness of Life: Biology and the Anthropic Principle

So far the anthropic principle has mostly yielded statements about cosmology and astrophysics. The arguments employed properties and requirements of life on Earth. Amazingly enough, there is also the converse possibility, that is, to start from astrophysics and apply statements from that to the question of life on Earth. Such a direction of argument seems new and went rather unnoticed by evolutionary biologists. At a conference in London held in 1983 British astronomer Brandon Carter (Observatoire de Paris at Meudon) presented such an argument. In essence he claimed that on the basis of the anthropic principle one could predict that exterrestrial civilizations should be extremely rare — or that life on Earth could even be a unique case.

To demonstrate his case Carter starts from an obvious property of biological evolution of life on Earth which had neither been noticed much nor recognized as a further coincidence. "What impressed me," says the astrophysicist, "is the fact that it did take us almost as long to develop on this planet as the time for which the sun is available as an energy source."

On the one hand, astrophysical widsom has it that the sun will stay in its present state only for about ten billion years. For biological evolution to take place on a planet around the sun, this is the time available for a suitable ecosphere. After this time interval, the sun will expand and

197

swell up and turn into a "red giant" and thereby extinguish any life left on Earth.

On the other hand, the evolution of life has taken almost four billion years to produce Man — using the available timespan of the sun within a factor of two. Hence there is a remarkable coincidence between life on Earth and our fixed star: *The time available for life on Earth equals — within a factor of two — the time which it took to produce intelligence.*

"I consider this fact extremely important," comments Carter. "I believe that it can be used to prefer certain models of biological evolution. In particular it contradicts the popular view held by many science-fiction authors that the universe teems with highly-developed life."

It is assumed that chance events also influence the process of evolution, among them certainly a number of highly improbable events. If one tries to estimate the chances for the rise of a civilization with a given ecosphere on the basis of a simple statistical model, two possiblities arise to explain the above coincidence:

Possibility I: The average time for producing intelligence by evolution is short when compared with the astronomically available time (ten billion years — in case of the sun). Or

Possibility II: This time interval is much longer than the astronomically available time.

If the first choice were true, then there should be almost as many civilizations in the universe as there are ecospheres. Then, however, it would be hard to understand why life on Earth — when seen against the solar lifetime — arose so slowly and comparatively late. If, on the other hand, the second alternative were correct, then even in the most favorable ecospheres intelligence would not originate; their stars would die too soon. Intelligence would evolve only late within the given time interval—if at all. Using the anthropic principle now as a selection principle, the Earth would constitute such a rare case.

Although this argument of Carter's is not really compelling, one is led to prefer possibility II for the simple reason that it better matches the data available — i.e., the present lack of evidence for extraterrestrial intelligence. Continuing the argument, if many quite improbable events did play a role in the evolutionary process, then the average time for the evolution of intelligence would shift towards the latest possible moment in any available time interval. Then it would be quite a surprise as to why intelligence on Earth has arisen already, and not just before the sun turned red giant.

From this dilemma Carter concludes: "For me this seems to suggest that only one or two events of the entire evolution could have been highly improbable." Only in this case would the average time for the evolution of intelligence not be so special (and long) for many planets endowed with an ecosphere. Only then would there occasionally also be a "shorter" evolution up to the intelligent state (like on Earth). "To make it definite, this leads to the prediction that all attempts to find extraterrestrial life will surely fail."

INTELLIGENT LIFE FORMS AND THE PHYSICAL LAWS

One of our universe's noteworthy qualities is the existence of intelligent life forms that have developed instruments with which they in turn observe the universe. The existence of life is more or less directly "influenced" in two different ways by the fundamental constants that appear in the physical laws. The interplay of the natural forces provides suitable material and planetary surroundings; they make available the proper ingredients life needs for its development; and lastly they influence the evolutionary paths that made possible homo sapiens appearance on the cosmic stage.

Until now we have mostly examined the environmental conditions and the material ingredients which would have been missing in the universe if the fundamental constants would have had other values. The influence of special extra-terrestrial and global events — effects of cosmic catastrophes, for instance — on the paths of evolution and the creation of complex forms of life was discussed in the last chapter. How the construction and evolutionary development of genetic and biological systems depend on the natural constants shall form the keystone of our investigation of the anthropic principle. "Insofar as one can show that changes in these constants prohibit the existence of advanced life forms,

199

this is of importance in illustrating why such constants have particular value in 'cognizable' universes; and even if such changes do not make life impossible, they almost certainly must imply changes in the details of such life forms that exist," write G. F. R. Ellis and J. Kreuzer of the University of Cape Town. The phenomenon of evolution is separate from the mechanism of evolution. Darwin only attributes a main role in evolutionary events to the process of natural selection. But catastrophes of all kinds and a change in the natural constants, if it is possible, can also hasten the development of the species. In the discussion of the anthropic principle up to this point we have dispensed with trying to define more precisely what life itself or an intelligent life form is.

Decades of discussion indicate that the phenomenon "life" is difficult to differentiate from inanimate matter, principally because the transition point between animate and inanimate matter is vague. In order not to have to deal with assorted virus-like, quasi-crystalline transitional forms, it suffices for our purposes to examine only intelligent, biological life forms that serve themselves with complex technical tools. In this way it may be uniquely decided in each case what life is and what it is not. We will examine in the next section how forms of life can be comprehended purely structurally as "generalized living systems." The point here of importance in connection with the anthropic principle concerns the places where biological function can be linked directly, not arbitrarily, to the physical laws. The connection between biological function and the natural constants is especially clear in the following places:

— in the role of water for the existence of life;
— in the nature of the chemical bonds for genetic material;
— in the function of the nutrients important to life, the trace elements which include several of the very rare elements;
— in the effect of approximately 2000 enzymes (proteins which work as bio-catalysts) in genetic reactions.

LIVING SYSTEMS — A 7 BY 19 MATRIX?

Over the last several years, although not known to the general public, the discipline called systems theory has identified the important qualities of living systems and has organized their characteristics into different levels.

One characteristic feature of biological evolution is that over the course of their development their systems become increasingly more complex. Biological systems regulate themselves and in so doing optimize certain functions. With the propensity for greater complexity comes, among other things, the fact that living systems represent a state of non-equilibrium which must be maintained through a constant supply of outside (free) energy against the destructive influences of the environment. Because of this, at every point in time there is a multiplicity of life-forms, from the simplest to the most complex, which are simultaneously capable of life. One interesting thing systems theory has demonstrated is that although the various levels of living organisms differ in complexity, they all fulfill practically the same categories of well-defined functions. These categories are extremely helpful in differentiating between animate and inanimate matter!

Every form of life reveals itself to be built up of several hierarchically graduated levels. In each case these levels regulate themselves as much as possible, but they are in communication through commands given from the bottom to the top and vice versa. Every level provides for itself to a certain extent, regulates its own energy needs, and protects itself in some way from the environment. The American systems theorist James G. Miller identified seven levels which extend from the microscopic to the socio-political stage that are layered over one another while at the same time being embedded in each other: cell, organ, organism, group, organization, society, and supranational system.

What differentiates the individual levels from each other — aside from the size of the systems — is the circumstance that with every higher level new characteristics and structures appear, resulting in an organizational advantage. "These enable [the systems] to cope with excess and lack stresses that would be beyond the adjustment capabilities of lower-level systems, and these characteristics also result in greater structural complexity than exists at lower levels," says Miller. The capabilities at every system level are essentially the result of cooperation of the many elements of a lower stage. "At the level of the cell, the emergent of course is life itself — a new sort of organization of matter and energy that can maintain an island of negative entropy and stability in an environment with less stability and a greater overall rate of entropy." At the level of the organ, the lifespan, for example, rises in comparison to cells. "The ability of an organized mass of cells to replace its constituent cells as they die is emergent at the level of the organ. As a consequence, many types of organs live longer than the individual cells that compose them." New forms of cooperative activities and qualities come about also in the levels of social groups and organizations: language, writing, the exchange of money, personal status, bureaucracy, and so on.

Only the three lowest system levels — cell, organ, and organism — belong solely in the realm of biology. Of these levels it can already be said that the cell can care for itself in large measure, since single-celled creatures (protozoan) are capable of existing alone. If cells could not provide for themselves, then this function would have to be taken over by the higher organs which would naturally be made up of cells which would themselves require maintenance. Therefore, the self-regulation of units is already essential at the lowest level of organization. At the next-higher system level, the organism, various units (brain) control the roles of the individual organs (limbs, senses). Conversely, the lower levels of the total system can send information about needs (hunger, thirst, fatigue) to the higher levels. In a bureaucracy this would correspond to the standard procedure of allowing decisions to be made at the appropriate lowest level at which they generally can be made. Only if a problem cannot be solved at one level is it pushed up "one higher." Systems with the greatest possible autonomy of single levels within a hierarchy obviously form an essential component of the construction of such complicated organisms as Man. The first step toward cells was already a decisive one, as George G. Simpson points out in his book, *The Meaning of Evolution*: "Above the molecular, the simplest fully living unit is almost incredibly complex... All the essential problems of living organisms are already solved in the one-celled protozoan ... The change from replicating molecule to protozoan was probably the most complex that has occurred in evolution, and it may well have taken as long as the change from protozoan to Man."

SELF-REGULATION AND HIERARCHICAL CONTROL

The greatest possible self-governing of all the system levels finally made possible the formation of an organism, formed from 60 million million (6 $\times 10^{13}$) single cells, directed at the highest level, the brain, whose control function we are now only beginning to understand. This is an entirely

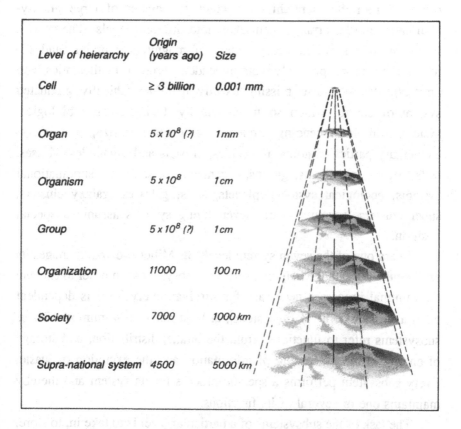

Level of heierarchy	Origin (years ago)	Size
Cell	≥ 3 billion	0.001 mm
Organ	5×10^8 (?)	1 mm
Organism	5×10^8	1 cm
Group	5×10^8 (?)	1 cm
Organization	11000	100 m
Society	7000	1000 km
Supra-national system	4500	5000 km

The seven hierarchical levels of living systems exhibit within each level the same internal organization. This differentiates them from inanimate systems.

203

different kind of structure from that of the universe. Frequently, planets, stars, galaxies, and clusters of galaxies are viewed as a form of "organism," as a life-form of its own kind, that is to say, as the cosmic continuation of the biological hierarchy on Earth. In fact, some do not balk at addressing the entire cosmos as a form of life of special dimension that also stays in contact with similar cosmic life-forms — whatever such a clash of words may mean. (See for example W. Bargatzky's *The Cosmos Lives* and other books at a similar level.) At the base of this association, there lies a fundamental misunderstanding, which can be removed by the system theoretical view. The hierarchical control effective in biological systems is completely different from the hierarchies by which the matter of large gravity-dominated entities arrange themselves into different levels. The cosmic forms of matter are not only orders of magnitude simpler (and for that reason more precisely comprehended through mathematics and physics), they are also missing exactly that hierarchically graduated system of control which so fundamentally distinguishes a biological system. In the ascending sequence of levels (by size, that is) — elementary particles, atoms, molecules, crystals and organelles, viruses, cells, organs, organisms, groups, organizations, societies, supranational systems, ecological systems, planets, stars, galaxies, galaxy clusters, super clusters, cosmos — the seven living systems assume a special position.

Every one of the seven system levels, as Miller showed, manages its self-regulation with the help of a series of subsystems. In order to execute the minimal processes necessary for survival, every level is dependent upon a series of critical subsystems, at least nineteen in number. These subsystems refer to functions: from the intake, distribution, and storage of energy to the processing of information and the excretion of waste. Every subsystem performs a specific process for its system and thereby maintains one or several of its functions.

The task of the subsystems of a particular level is to take in, to store, or to process matter and other forms of energy, to react in an adequate fashion to the influx of information and to give out information signals about itself in turn. Of the 19 critical subsystems only two are engaged in processing matter and information — the reproducer system and the system's boundary. The *reproducer system* takes care that further systems of the same type come about. This happens at the most basic level in the cells — egg and sperm cells — through mitosis. Through reproduction not only is the blueprint of the cell system — contained in the DNA of the chromosomes — further extended, but the matter for the construction of the new life system is organized. The "child" system is looked after and provided for long enough for it to become independent and provide for itself. The *boundary* protects the system from the toxic influences of the environment and controls the influx of selected matter, energy and information.

A series of other subsystems are concerned solely with the processing of matter, especially energy. The *ingestor system* is one of them. It conducts materials from the environment over the border of the system. After that a *distributor* transports the material brought in from the outside (the "input") or substances from other subsystems to components in the entire system. Miller states, "a sponge has its components so arranged that they form canals into and out of which sea water can freely flow. They form its distributor... groups often constitute a member that concerns itself with food or other provisions. In higher organizational systems these are special groups, which, with the help of auxilliary means such as streets, trucks, railroads, and airplanes, represent this subsystem." The *converter* comes into action when there are many input-substances that must be prepared in order to be able to find an application in the special processes of the system: what the enzyme is in the cell, the intestine in the organism, the cook in the family, and the factory in society. The *producer* produces material that is necessary for growth, repair, or substitution. In the cell this corresponds to chemical synthesis, in the organism to the functions of the liver, bone marrow, and the like, in society approximately to the function of the pharmaceutical industry. The *matter-(energy) storage system* preserves materials in time, after which they make themselves available again to the system and before which they, after use, are removed from the system as waste by the *extruder*.

Finally, the *motor* and the *supporter* provide for the spatial mobility of a system in regard to the environment, and also avoid overload (or overpopulation) of the system (emigration).

The bureaucracy of the systems, so to speak, is represented through the third group of critical subsystems. They only process information and look after meaningful reactions to stimuli of the outer and inner world

THE HIERARCHY OF LIVING SYSTEMS: a 7-by-19 matrix in examples

According to the theory of living systems, every form of life maintains its function with the help of 19 subsystems. These occur in all living systems and are given in the 19 horizontal rows of the tables below. The seven different levels of living systems are given in the columns. Each box offers only one selected example of a subsystem at the respective level. In three cases, the process appears to be dispersed to another level.

TABLE 1

Subsystems processing both matter (energy) and information

Level ——— Subsystem	Cell	Organ	Organism
Reproducer	Chromosome	None; downwardly dispersed to cell level	Genitalia
Boundary	Cell membrane	Capsule of viscus	Skin

206

which are necessary for survival and stability. In their functioning all the concepts that information theory, communications-channel theory, and cybernetics describe come together: the processes of learning, remembrance, forgetting, adjustment, ignoring, action, and flight.

The astounding thing is that in every one of the seven system levels all the critical subsystems meet the identical tasks almost down the line.

Most unknown components occur in the subsystems responsible for association and memory, although it is clear that these functions are carried out. These principally concern information processing at cell, organ and organism level.

Table 1: Two subsystems process matter (energy) as well as information.

Table 2: Eight subsystems process matter (energy) only.

Table 3: Nine subsystems process information only.

(Tables after J. G. Miller: *Living Systems*, McGraw-Hill, 1978)

Group	Organization	Society	Supranational Systems
Mating dyad	Group which produces a charter for an organization	Constitutional; convention	Supranational system which creates another supranational system (U.N.O.)
Sergeant at arms	Guard of an organization's property	Organization of border guards	Supranational organization of border guards (U.N. Peace Corps)

Occasionally one of the functions is displaced at a deeper system level and sometimes one is still unknown (see Table). The unliving cosmic entities from planet upward fail to have comparable self-regulation as well as the control hierarchy between individual levels of the organizations.

Now, in order to fix the connection between biological systems and the fundamental constants we must identify the biochemical basis for all life-forms. All the immensely complicated biological systems are controlled, strictly speaking, by one particular molecule contained in the hereditary material of cells: dioxyribonucleic acid (DNA). It is in the qualities of this hereditary molecule, made from a helical double strand, that we can recognize the effects of the physical laws. They set the type of chemical bond between single atoms of the DNA molecule and therefore also determine the possible spatial arrangement of it. They naturally also fix the structure of all the other molecules. Special organic molecules work in biochemical reaction principally with the aid of their "spatial configuration" or "spatial folding," the way in which the atoms or certain atom groups of every molecule appear to each other. The molecules in the genes often contain many molecular subgroups, which, in the framework of the entire molecule can turn relative to each other like the joint of a finger. They assume certain angles in relation to each other which minimize the forces of electrostatic repulsion. In simple cases, the minimizing of the repulsive forces creates "bulges," "buckling," and "concavities" in molecules through such folding, from which the chemical activity of the molecule starts. A bulge, for example, can fit into the molecular concavity of another molecule like a key into a lock. The molecular structure of DNA "is governed by the bond lengths and angles which are governed in turn by the quantum mechanical laws which determine the nature of the chemical bond. Regardless of whether life is strictly mechanistic or not, this physical structure, by which the function of the body is controlled, must be correctly maintained, or the functioning will be either impaired or impossible. Putting these features together, then, we do not have to understand all the links in the chain by which the DNA molecule and enzymes control biological structure and function to note that if the laws of physics were substantially different, the possible molecular configurations would be different" (Kreuzer/Ellis).

In biology it is often not so easy to determine in individual cases how much certain qualities of a substance or the reactivity of a molecule would change if the natural constants were altered. In contrast to the construction of a star, for instance, which would barely change if the gravitational constants were adjusted, some qualities of biological systems would change radically. Still, we shall attempt this for several essential points — with water and the rare elements — in what follows.

Water is Best!

The existence of life depends in many different ways on water and its qualities. Water's role is central — as solvent, as nutrient transporter, and as a partner in chemical reactions. It is water that we predominantly consist of: Man contains 60 percent water, 80 percent of the blood is water, the brain is quantitatively a damp sponge with 70 to 75 percent water, and even the dry bones are still 20 percent filled with the noble drink. Some algae and jelly-fish contain only two percent solid material. Water is an essential constituent of our planet's surface through which it became completely indispensable for the evolution of life. Life adapted itself optimally to the qualities of water. (Alternative biological evolution in other liquids has not yet been observed.) In any case, in water all the chemical reactions took place which led to the formation of the first primitive single cell organisms. Almost all the chemical reactions in the interior of living cells proceed in watery surroundings. In conclusion, as the textbooks would put it, there is "sufficient indication" that "water is an active participant in many biochemical reactions and that it essentially determines many of the qualities of macromolecules, like the proteins." Doubtless water, like blood, is "a most singular juice," a fundamental component of plants, animals, and humans, as well as the total ecological system of the Earth's surface.

No other molecule has so central a role in terrestrial life as H_2O, which should not imply that it is irreplaceable; on other planets under exotic cosmic environmental conditions other essential fluids may serve the same purpose. However, on Earth nothing can substitute for water! The reason for this lies with several very simple physical and chemical

209

TABLE 2

Subsystems processing matter (energy)

Level / Subsystem	Cell	Organ	Organism
Ingestor	Gap in cell membrane	Input artery of organ	Mouth
Distributor	Endoplasmic reticulum	blood vessels of organ	Vascular system
Converter	Enzyme in mitochondrion	Parenchymal cell	Upper gastro-intestinal tract
Producer	Enzyme in mitochondrion	Parenchymal cell	*Unknown*
Matter (Energy) Storage	Adenosine triphosphate (ATP)	Intercellular fluid	Fatty tissues
Extruder	Gap in cell membrane	Output vein of organ	Urethra
Motor	Microtubule	Muscle tissue of organ	Muscle of legs
Supporter	Microtubule	Stroma	Skeleton

Group	Organization	Society	Supranational System
Refreshment chairman	Receiving department	Import company	Supranational system officials who operate international ports
Mother serving food to family	Driver	Transportation company	U.N. Children's Fund (UNICEF), providing food to needy children
Butcher	Oil refinery operating group	Oil refinery	EURATOM, concerned with conversion of atomic energy
Cook	Factory production unit	Factory	World Health Organization (WHO)
Family member who stores food	Stock-room operating group	Warehousing company	International Red Cross which stores materials for disaster relief
Char	Delivery department	Export company	Component of the International Atomic Energy Agency (IAEA) concerned with waste disposal
None; laterally dispersed to all members of group, who move jointly	Crew of machine moving personnel	Transport company	Transport component of NATO
Person who physically supports others in group	Group operating organization's building	National officials operating public buildings and land	Supranational officials operating U.N. buildings and land

Group	Organization	Society	Supranational System
Members of group signalling to other members	Private telephone exchange	National telephone	Universal Postal Union (UPU)
Interpreter	Foreign language translation group	Translation unit	Supranational translation unit
None; laterally dispersed to members, who associate for the group	*None*; downwardly dispersed to individuals (organism level)	School, university	Supranational
Adult in a family	Filing department	Library	U. N. Library
Head of a family	Executive office	Government	Council of Ministers of the European Communities
Composer of group statement	Speech-writing department	Press secretary	U.N. Office of Public Information
Lookout	Telephone operator group	Foreign news service	News service informing supranational system
Group member reporting group states to decider	Inspection unit	Public opinion research	Supranational inspection organization
Spokesman	Public relations department	Office of government spokesman	Official spokesman of Warsaw Treaty Organization

212

TABLE 3 *Subsystems processing information*

Level / Subsystem	Cell	Organ	Organism
Channels and Networks	Cell membrane	Nerve net of organ	Components of neural network
Decoder	Molecular binding site	Receptor cell of sense organ	Cells in sensory nuclei
Associator	*Unknown*	*Unknown*	*Unknown*
Memory (Data Storage)	*Unknown*	*Unknown*	*Unknown*
Decider	Regulator gene	Sympathetic fibre of sinoatrial node of heart	Part of cerebral cortex
Encoder	Hormone producer	Presynaptic region of output neuron of organ	Temporoparietal area of dominant hemisphere of human brain
Input Transducer	Receptor site in cell membrane	Receptor cell of sense organ	External sense of organ
Internal Transducer	Repressor molecule	Cell of sinoatrial node of heart	Receptor cell responding to changes in blood states
Output Transducer	Presynaptic membrane	Presynaptic region of output neuron of organ	Larynx

213

qualities of the water molecule. The two hydrogen atoms (H) form with the single oxygen atom (O) a triangular molecule (angle at the oxygen atom: 104.5 degrees) whose electrical charge is unequally distributed. Water is "polar": the O-side is negative, the opposite side with the two H atoms is positively charged; this turns the molecule into an electrical dipole, similar to a small bar magnet with oppositely charged ends.

This inconspicuous physical detail has dramatic consequences; it makes water universally operational —chemically and biochemically:

— Electrically charged molecules (ions) attract water molecules, wrap themselves in a coat of water and in that way become water soluble in the same way (saline state) and can move almost freely — as soon as they lose their water wrapping, they become bound up again as colloids. Osmosis and the permeability of cells hold together because of it.

— Water transports nutrients and wastes in plant and animal cells.

— Water takes part in photosynthesis. Through the effects of sunlight, carbon dioxide and water are transformed into energy-rich hydrocarbons — the chemical storage of solar energy. In the first phase of this, "photolysis," the splitting through the effect of light, plays a role. In photolysis, water is split by the radiant energy of the sun into what are known as radicals.

— Water has a high surface tension, which is an impediment to dishwashing without detergent. But it supports splendidly the formation of cell membranes and protein layers.

— The surface tension assures that in plants and trees nutrients are brought through meter-long canals and ducts up to the leaves.

— Water can absorb a great amount of heat (it has a high "thermal capacity"); because of this a cell full of water warms up only negligibly, even when heat-generating (exothermic) reactions take place.

— In order for water to evaporate, a relatively large amount of heat is needed, so biological systems are able to effectively cool themselves through evaporation.

Also, water is out of the ordinary in other kinds of physical behavior — again with biological consequences. When water freezes it expands. Therefore ice forms on top of water and not at the bottom of lakes and oceans which helps the marine world survive winter. The importance of this for flora and fauna is obvious. At 40°F water is at its densest and at at 115° it is most difficult to compress. It melts and evaporates at temperatures which for a non-metallic substance formed from light atoms is unusually high. Water becomes more "fluid" if it is put under high pressure. Several of these qualities are also found in other substances, "yet, it is striking that so many eccentricities should occur together in one substance," concludes water researcher Frank H. Stillinger.

The angular form of the water molecule is determined mainly by the mutual electrical repulsion of the eight outer electrons of the one oxygen and two hydrogen atoms. These form four electron pairs, two "free" pairs and two bound to the hydrogen atoms which arrange themselves in the four corners of a tetrahedron. If the electrical force, specifically the electromagnetic fine structure constant, had a different value, the distance of the atoms in the H_2O triangle would change, as would the chemical bond, the angle arrangement, the spatial effect and the polarity. While it is difficult to make quantitative statements, it is also clear that the electrical force could only change within specific boundaries if the multiplicity of functions which water has assumed in the organization of life is not to be endangered. Of course it must remain an open question whether or not with such altered fundamental constants other connections could take the place of water.

A WATER SUBSTITUTE?

Among other functions, the sun has kept the Earth over a long period of time at stable temperatures somewhere between the freezing point and the boiling point of water. Neither on Mars nor on Venus, which is otherwise so similar to Earth, is there a quantity of water in liquid form worth mentioning. In neither the Martian ice deserts nor the Venusian sulfuric acid steam oven could the evolution of life have been able to get beyond the pre-biological phase. Water, indeed irreplaceable for

215

terrestrial life, could perhaps under other circumstances, on other planets, be replaced by other liquids. But it is not easy to find a liquid which can substitute for water. Even *heavy* water (D_2O), formed with the hydrogen isotope deuterium, exhibits different behavior; it freezes at a slightly higher temperature. On the cosmic scale, heavy water is ten thousand times rarer than normal water. Other potential replacements — such as methane (CH_4), hydrogen sulfide (H_2S), carbon dioxide (CO_2), hydrogen fluoride (HF), or sulfur dioxide (SO_2) — all melt below minus 50 degrees Celsius and are at most one tenth as "polar" as water.

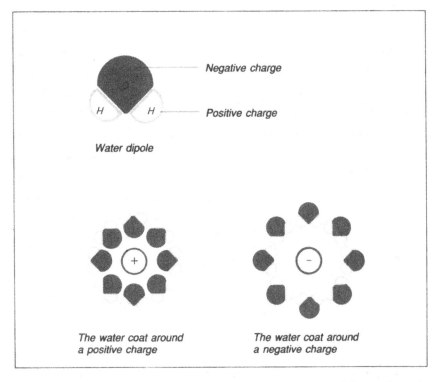

Negative charge

Positive charge

Water dipole

The water coat around
a positive charge

The water coat around
a negative charge

Water, the basic elixir of our life form, has one special quality. Its molecule has both a positively and a negatively charged end. On account of this, ions are easily water soluble, as they surround themselves with a "coat of water."

If on other planets, although not on Earth, different substances could possibly take over the role of water, SO_2 and HF might be good candidates. But at 900 degrees it is too hot on Venus, whose clouds contain sulfuric acid vapor, for oceans to be filled with sulfuric acid. On Mars temperatures are suitable for HF; but only water has been found, in the form of ice. (In 1980, radar measurements provided indications of seas of liquid water under the surface of Mars.)

Of all the alternative candidates, ammonia comes out the best. Although the ammonia molecule is five times less polarizable than the water molecule, ammonia is similar to water on several points. Ammonia, NH_3, which exists as a gas at room temperature on Earth, has almost the same atomic weight (17) as water (18); However, at atmospheric pressure it condenses at -27 degrees and freezes at -108 degrees. Ammonia dissolves foreign substances almost as well as water and requires a similar amount of heat to evaporate. A planet like Jupiter, however, would be too cold for ammonia. Under higher atmospheric pressure its melting and boiling points would be pushed to higher temperatures. Venus, as already mentioned, would be too hot and unfortunately ammonia doesn't exist there anyway...

Could biological systems develop in liquid ammonia? What brought about the evolution of proteins over billions of years in water on the Earth should not be inconceivable in ammonia somewhere else. As long as the cosmic and material requirements are fulfilled, altered fundamental constants could favor different planetary regions, different life promoting zones, and perhaps even different substances than this peculiar, polar solvent we call water. Proteins and nucleic acids are gigantic molecules "which allow for an almost endless multitude of structures and therefore possess the variety which they require as the foundation of nearly endless variations of life forms," wrote science fiction author and biochemist Isaac Asimov optimistically in 1963. We know today that biological molecules can best fulfill their specific function in the network of the genetic code only with the most optimal spatial arrangement. Experiments have shown that alterations of even seemingly insignificant molecular details weaken the functioning of the entire molecule and with it the survivability of the life form which they serve. It appears that despite its great variety, the molecular biology of the cell operates with the precision of microscopic clockwork; even trivial changes in the structure of the molecule drastically interrupt its effectiveness. In the next section we want to examine this "spatial specificity" more closely.

Altering the electromagnetic fine structure constant affects in all molecules (not just water):
— the spatial construction (the angular arrangement in chains of molecules) and
— the strength of the chemical bond.
Next to the forces of the electrical charges, quantum mechanics plays a fundamental role in the chemical binding of atoms to each other. Through its microphysical laws it clarifies the deformation of atomic nuclei and the shell-like construction of the electron clouds which surrounds them. Naturally, these laws also fix the behavior of electrons, when, coming from different atoms, they get into molecular combination with each other. The fundamental quantity of quantum mechanics is Planck's constant (h). It is related to the electromagnetic constant by: $(\text{alpha}-E) = \frac{2\pi e^2}{hc} = \frac{1}{137}$ (e is the electric charge of an electron, c the speed of light.)

Due to its simplicity the hydrogen atom best demonstrates how its structure is controlled by alpha$-$E. Every molecular structure contracts in size if alpha$-$E increases because the electrons would then be more likely to be closer to the atomic nucleus. (This is somewhat analogous to the planetary orbits around the sun with stronger gravity; the planets would orbit more tightly around the sun.)

With the large biomolecules, the proteins and enzymes, chemical functioning is determined by their "folded structure." The molecular chains ("polypeptide chains") of proteins can arrange themselves into levels crumpled into each other, which look like a piece of paper folded into zigzags. Another possibility takes the form of a simple spiral staircase, the alpha-helix. In it the molecule winds itself in the form of a spiral, which at specific points is "cross-braced" by hydrogen

218

bonding. In the majority of proteins several twisted and stretched molecule chains wind around each other — they form the so-called "tertiary" and "quartemary" structure of protein. The three-dimensional folding creates active centers in the molecular hollows and bendings. The German biochemist Manfred Eigen delineates the connection between spatial construction and function: "the specificity of substrate recognition, which is executed by the active centers, follows from the nature and exact spatial position of the side chains and their amino acids." A change in the folding also destroys the particular function because "it seems clear, that these molecular control mechanisms are critically dependent on the structural details of the catalytic and regulatory polypeptide chains...Not only is the quality critically dependent on the structural details of the catalytic and regulatory polypeptide chains... but the quality and variety, as well as the homeostatic balance of cell metabolism are all determined by the amino acid sequences of the protein molecule."

The dtv-Dictionary of Biology describes the spatial construction of DNA: "A DNA molecule consists of two unbranched threads of polynucleotides, which as a right-turning double helix runs along a common axis. The bases (molecular building blocks) stand at right angles to the axis pointing inward, whereby they always lie opposite complementary molecules and form hydrogen bonds. The two strands are set spatially across from each other such that on the DNA surface a deep and a flat groove is formed."

The physical laws underlie the specific folding of biomolecules. If the constant alpha–E changed, then every angle would change, and the chemical activity centers would lose their effectiveness. The key would no longer fit in the lock. Perhaps other centers would then develop chemical activity and the molecular construction would be flexible enough to form analogous structures. But that remains pure speculation. One could equally expect the molecules to simply become chemically functionless.

No Copper, no Vanadium — no Life!

In the same way, rare elements and heavy atoms, as trace elements play an important role for life; with their very special qualities they are, in a manner of speaking, the "spice" in the soup of life. In addition to the typical elements of organic material — hydrogen, carbon, nitrogen, oxygen, sulfur, phosphorous — organisms need mineral elements, especially mineral salts. In contrast to organic material these are neither produced nor consumed by the organism. They also serve as building blocks in protein combinations important to life. Magnesium is an example: magnesium atoms are components of the enzymes without which photosynthesis could not proceed.

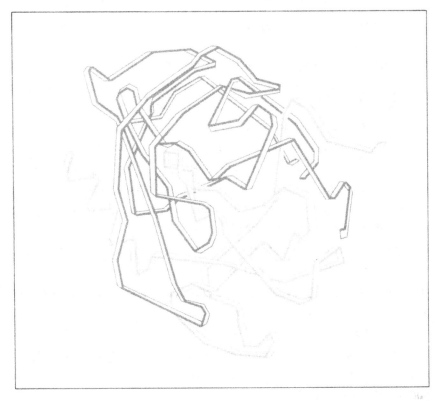

The biochemical effect of an enzyme (biocatalyst) depends on the spatial folding of its molecular chain. Altered fundamental constants would modify this folding and with it the molecule's effect. (After Eigen, 1971, p. 498.)

220

To the elements important to life belong:
— the macro-elements calcium, magnesium, sodium, potassium, phosphorous, chlorine, and sulfur; they are all considered "essential";
— the trace elements manganese, cobalt, iron, molybdenum, copper, and zinc, as well as selenium, iodine, chromium, tin, fluorine, nickel, vanadium, and silicon. The first six in the group are essential as micronutrients in freshwater ecosystems.

If the chemical qualities of one of these nutrients is changed, it is to be expected that their effective metabolic function will break down or at least be disrupted. If the feeding system is "disrupted" by changes in the fundamental constants — most effectively in the electromagnetic fine structure constant alpha–E — then gradually the entire system will be increasingly damaged. This can go so far that a specific change in the natural constants, which modifies the chemical function of these elements, can prevent or drastically transform the creation and development of life. Most severely affected by this would be copper, nickel and vanadium. Iron is also an essential component of the hemoglobin molecule in blood, and copper is a part of two enzymes in the chemical chain reaction of oxygen respiration.

Many of the qualities of the electron clouds responsible for the chemical behavior of these elements depend directly on alpha–E; and indeed it is the relativistic effects that influence the construction of the outer layers of electrons, above all the electrons of the outermost shell. Variations in alpha–E would presumably alter the chemistry of the trace elements vital to life. Then other elements would have to be able to take over these roles in order to maintain their total function. Kreuzer and Ellis by reason of this consideration reach the noteworthy conclusion that the effects of "relativistic quantum theory are necessary for the viability of life." It comes into question here whether other elements could substitute in the subtle and so enormously complex networks of biochemical reactions, and whether such a different system would produce comparable achievements on which an intelligent life form is ultimately dependent.

At the close of this book therefore we want to put forward the hypothesis that the world must be essentially as it is in order to bring about a life form comparable to *homo sapiens*. The discussion of the anthropic principle as it has been conducted from Chapter III to VIII has furnished the first clues of support for this thesis. However, they can only be understood as a beginning of a discussion and the start of a more in depth analysis. The structural and evolutionary preconditions of intelligent life are known to us only in broad strokes. For a serious "proof" of this above mentioned hypothesis there is still much to do. For example, we must ask whether we reach this result because we know too little to be able to imagine meaningful, alternative, perhaps simpler "self-cognizant worlds." In the last chapter we take up this question.

Chapter IX
The Cosmic Life Plan

"Let us take a particular cosmological model as a basis for studying what happens when the constants of nature are changed in value. In this way we shall generate not just a new range of models of our universe, but a range or ensemble of universes corresponding to the given cosmological model. An ensemble of universes, covering all values of the constants of nature, apparently contains only a very small subset of universes in which stars and planetary systems exist. This leads to the conclusion: the constants of nature have their observed value because our universe is perhaps the only universe in which they can be observed by intelligent life."

With these words E. R. Harrison of the University of Massachusetts, Amherst, begins his discussion of various *principles* which modern cosmology uses as a hypothetical basis in order to filter out some meaningful interpretations from the wealth of data available. We, of course, are interested in the anthropic principle, which Harrison's study — taking up Dicke's idea — places alongside other cosmological principles for the first time though still under the name of "Principle of Cognizability." However, Harrison was under no illusion that this principle of "the biological selection of natural constants" could serve as the basis for a scientific theory: it is concerned with metaphysics and not with physics, whereas to be scientific a hypothesis, including these "principles," must be open to refutation. But how are we ever to examine and possibly refute the suggestion that the existence of many universes may be the reality and not just an invention of science fiction? In the field of literature, the concept of numerous "parallel" universes has been widespread for quite some time. In 1937, long before physicists began to dabble in such ideas, Olaf Stapledon (1886-1950) wrote in his novel *Star Maker* of the designer and creator of a multitude of universes, each cosmos inhabited by a particular form of intelligence. In 1972, thirty-five years after Stapledon's Star Maker, Isaac Asimov returned to the theme from a different angle. In *The Gods Themselves*, he conceives of a chain of universes in which entropy plays an important role. In each universe entropy grows as the sum of all processes taking place in it, but at a different rate from one to the next. An advanced cosmos in this scheme of things defers its over-hasty progress towards an entropy-induced icy end by tapping energy from a neighboring cosmos.

But back to science! Physical principles are employed for the purpose of understanding the structure of the universe. They are complemented by the symmetries, among which are homogeneity and isotropy, which we discussed in Chapter IV. According to Ellis and Harrison, "the symmetry principles provide a descriptive framework in which the physical principles provide the explanations."

Although we can make no physical examination of an "other universe" any more than we can prove or refute its existence, there is nothing to stop us from studying the variety of cosmological models we can construct by changing the natural constants. One example of these follows.

AN EXAMPLE OF AN "ALTERNATIVE" UNIVERSE

We can start by considering whether the anthropic principle "explains" why gravity is so immensely weak. It is perplexing that all bodies for which gravitation plays a role have a mass that equals the mass of a proton multiplied by a power of 1/alpha−G (see Chapter VI). If the connection is so simple, is it not possible to imagine a hypothetical universe in which all the laws of microphysics are unchanged but gravity, for example, is a million times stronger? What would this universe look like?

224

It would be a strange universe to our eyes, with the following characteristics as outlined by Carr and Rees. The masses of the stars and planets would be smaller by a factor of a billion and their diameters by a factor of one thousand (the Earth would have a diameter of 13 kilometers), and the life expectancy of the sun would shrink from ten billion to one hundred years. Observers in this speeded-up universe would find Dirac's cosmic coincidence fulfilled a mere hundred years after the Big Bang — although this alternative universe would contain a thousand billion times fewer particles than our own. The modified Big Bang would not have lasted as long and thus would have provided far less time for the production of particles.

But would observers, intelligent life, actually appear? Would this actually be a "self-cognizant" universe? It would seem to be the extreme difference between the natural forces which makes evolution and the formation of structures possible. The wide range of strengths of these forces may also be the reason why organisms are able to grow to a certain size before gravity dominates the chemical forces. A universe with a relatively small gravitational force contains many more particles, and thus contains more stars and finally more places in which a process of biological evolution can get under way. However, these relatively vague and qualitative arguments do not bring us closer to any numerical determination of the strength of gravity. In its place, on the other hand, we might offer the requirement that there must be sunlike stars.

But we do not need to raise the strength of gravity so drastically to see that these alternative universes would not be suited to the evolution of life. An increase of only a factor of ten is enough! Gravitation is 10^{40} times weaker than the electromagnetic force. Let us suppose that gravitation was just ten times stronger, and thus 10^{39} times weaker than the electromagnetic force. With just this tiny adjustment, a star like the sun would find its life expectancy sharply reduced: it would die out after only a million years — too short a time for biological evolution on Earth.

And a gravitational force only ten times weaker (10^{41} times weaker than the electromagnetic force) is also a non-starter. It would then be uncertain whether stars and planets would be able to condense out of the intergalactic dust and clouds; and the few planets that might still achieve a successful birth would be unlikely to hold on to any atmosphere. Stars, too, would get into similar difficulties. With lower gravity the stars would be smaller, and the pressure of gravity in their interiors would not drive the temperature high enough for nuclear fusion reactions to get under way: the sun would be unable to shine. Conversely, too high a gravity could overheat the interiors of the stars; as a result they would explode or collapse under their own weight to form black holes.

The constancy with which the sun continues to shed its life-giving rays on Earth derives not least of all from a stable combination of all four fundamental forces. In Chapter V we restricted ourselves to a discussion of the conditions making possible supernovas and partially convective stars (such as our sun). The presence of the sun's vital warmth depends on a "cooperative effort" by all the forces of nature!

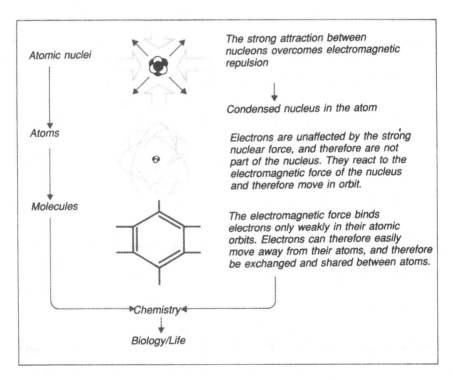

Life can only exist because the four fundamental forces of nature have different strengths.

In any "alternative universe" in which sunlike stars with long-term hydrogen burning are to exist — the phase in which the sun is now — all four forces of nature must interact in a specific way; it is only then that there is sufficient time for life to develop. To quote Freeman Dyson: "The energy flows on the Earth are embedded in the energy flows in the universe. A delicate balance among gravitation, nuclear reactions and radiation keeps the energy from flowing too fast." But why does the sun shine with such constancy and moderation?

Stellar energy is liberated by the nuclear burning of hydrogen to helium. More precisely: two neutrons and two protons must combine with one another to produce each nucleus of a helium atom. But for this to come about, half of all the protons must first be converted into neutrons — a very slow process, since it involves the weak interaction. The final step, the union of proton-neutron pairs (for deuterium) or nuclei with one proton and two neutrons (tritium), then takes place very quickly, as this is affected by the strong nuclear force. In the explosion of a hydrogen bomb, it is precisely this last step (and only this step) that is carried out, with deuterium and tritium used directly as the raw material. Thus, if it were not for the slow intermediate step of neutron enrichment, the speed of which is determined by the strength of the weak interaction, the sun would be a giant hydrogen bomb rather than a gentle slow burner. An astronomer in Bonn once remarked, "The peculiarity of cosmic energy extraction processes lies in their slowness."

But that is not the end of it. The Earth needs protection from the ten-million-degree heat of the hydrogen fusion smouldering in the center of the sun. The sun's opaque gas shell is the ideal "cloak" for this. Instead of flying freely out into space and reaching the Earth as an excess of heat, the radiation must slowly work its way up from the interior to the surface of the sun. As astronomer Michael P. Papagiannis writes: "It is as if Nature, like a loving mother, had dressed her children with warm, insulating coats. If the layers which surround the hot core were transparent, the sun would radiate like a body at ten million degrees instead of the mere six thousand degrees which is the present temperature of its glowing surface. Since the energy increases as the fourth power of

227

the temperature, the sun would radiate all its energy away in less than one day."

There is one more ingredient to the recipe. It may appear to be an accident that the sun's energy leaves its surface at a temperature of 6,000 degrees. But a body at 6,000 degrees radiates most energy at the wavelength of green light — and it is in green light that photosynthesis functions most efficiently in terrestrial plants. Clearly this is the result of a harmonization of plant life with the characteristics of sunlight through which an efficient form of chemical storage of solar energy was developed. If the electrical force binding electrons to the atomic nucleus had other values, sunlight would be either too weak to set photochemical reactions in motion or else too strong, thereby destroying the molecular structure of plant life or preventing its evolution from ever getting under way.

All four fundamental forces, then, combine to provide us with solar energy. Protons overcome the repulsive force between them only if they collide with each other at high velocities, for example in a gas; then the strong nuclear force unites them. The weak interaction provides the neutrons, and gravitation heats up the gas by means of the gravitational pressure of matter. Finally, the particles of light from nuclear fusion strike the atoms of the sun's gaseous shell — an electromagnetic effect — so that they only reach the Earth very slowly and at low temperatures, promoting photosynthesis on the surface of the planet.

PRINCIPLES IN COMPETITION

If the anthropic principle is to be brought into the cosmological discussion, this would be done in the context of the other principles employed in the search for scientifc explanations. The central pivot of *cosmological* principles is the aim of freeing the interpretation of astronomical observations from the Earthbound state of those who made them — that is, from the possibility that certain phenomena may be valid only for our particular position in space and time. Principles such as the anthropic principle, on the other hand, aim to make strategic use of the *special role* of the human race. This is a way of thinking that may well be implicitly taken for granted by biologists, anthropologists and paleontologists; but in physics it makes for a completely new approach, which was only introduced by the spiritual fathers of the anthropic principle.

"Physicists and astronomers tend to focus on the physical universe, from atoms to galaxies, and usually ignore the existence of Man. Philosophers and theologians, on the other hand, are preoccupied with Man and God and show very little interest in the material Cosmos. Man, however, is an integral part of the physical universe and therefore, in our efforts to comprehend the whole of creation, we must stop considering Man and nature as two separate and almost unrelated entities, and start searching for the true place and role of Man within the framework of the physical universe," states Papagiannis.

How the anthropic principle can be employed to enrich our store of knowledge was the subject of the preceding chapters. The "biological selection of natural constants," however, need not be a mere theoretical game with a variety of cosmological models. We can think of a concrete example even in the context of a cyclical cosmology. As we mentioned in Chapter IV, the fundamental constants could well have different values in consecutive cycles of the cosmos — with the result that life could exist in some cycles but not in others. It is possible that an "early" cycle — or the very first cycle of all — began with a relatively strong gravitational constant "as a zero-energy quantum fluctuation [of the free gravitational field]...involving only a few quanta," as Robert H. Dicke and his Princeton colleague Jim Peebles have speculated. With each collapse of the oscillating universe at the end of each consecutive cycle, a little more matter might be created in the cosmic soup, tapping energy from gravitation from cycle to cycle. Biological selection could then come into play when, in a particular cycle, the natural forces "accidentally" harmonize in such a way as to provide an opportunity for the origin of life: "Finally gravity is weak enough to permit stellar evolution at a slow rate that provides a hospitable environment long enough to permit biological evolution."

Besides the — still unconventional — anthropic principle, cosmologists make use of the following other principles:
- the Position Principle, stating that the likelihood that an observer will occupy a special or favored position (or special point in time) in the universe is vanishingly small;
- the Principle of Uniformity of Natural Laws;
- the Gravity Principle;
- Dirac's Principle;
- the Chaos Principle; and
- the Bootstrap Principle.

Let us look at these in order, beginning with the Position Principle. Although there are of course special locations in our cosmos, we cannot work on the assumption that we ourselves happen to occupy one of them. If there is anything special about our position, this must first be established in order to make us an exception to the Position Principle. (For example: the Earth is located in the interior of a galaxy; this galaxy is a member of a cluster.) With the anthropic principle we can contradict the Position Principle in its time dimension; regarded from the point of view of time, we are living in a special cosmic era (see Chapter IV).

While it follows from the Position Principle that the universe, regarded from some other point, looks exactly the same as it does from the Earth, it leaves open the possibility that the laws of nature vary from place to place or are dependent on time. This is where the second hypothesis, the Uniformity Principle, comes in. The basis for the claims made by the Uniformity Principle is to be found in observations of the sort made of quasars in examining the universality of the electrical fine structure constant (see Chapter VI): phenomena in a particular direction in space clearly obey the same laws as similar phenomena in a second direction that have no causal connection with the first. As the Position Principle states that this cannot be ascribed to the special position of the Earth, it must be concluded that the laws of nature are in all probability the same elsewhere as on the Earth. The uniformity of natural laws, then, can only be claimed when observations are combined with the Position Principle.

230

It also follows directly from the Uniformity Principle that the laws of nature have been valid without variation from the beginning of the cosmos. This considerably simplifies mathematical dealings with the Big Bang and its unphysical "singularity"; but some cosmologists find themselves troubled by the problem of how it came about that the physical universe began to exist right from the start in conformity with a particular set of physical laws.

The third hypothesis, the Gravity Principle, is probably the best known, for it contains a hypothesis that is usually taken for granted. Almost all familiar phenomena in the range between atoms and galaxies in size are ruled by electromagnetic and gravitational forces. Although the electromagnetic force is the stronger of the two, it is cancelled out over even short distances — which is why as a rule matter is electrically neutral. In the case of planets and stars, therefore, the otherwise very weak force of gravity outweighs the forces of the microcosm. Outside of the range just mentioned, other forces are involved — in the sub-atomic dimension, the nuclear forces. At the other end of the scale, in the supergalactic sphere, across the distances between clusters of galaxies and even between clusters of galactic clusters, it is generally believed that, unlike the world of particles, no additional forces are at work and gravity continues to regulate the interactions between objects. But is it possible that other forces come into play at these largest of cosmic distances? Harrison considered it "not inconceivable that in the future our ideas on the nature of space, time and gravity on the cosmic scale will be entirely different from current ideas."

Dirac's Principle was covered at length in Chapter IV. With its aid, Dirac looked for a physical explanation for the cosmic large-number coincidences, specifically by assuming that gravity declined with time. We have discussed these attempts fully in earlier chapters. In contrast to the anthropic principle, Dirac tries to provide a *physical* reason for the existence of these large numbers. Physicists may well seek in the future to find such an explanation for anthropic relationships. Similar considerations apply to the Chaos Principle in its concern with the primordial situation when the cosmos began. Many find it incomprehensible or improbable that the cosmos was extremely homogeneous and isotropic at a very early stage (see Chapter IV) and have sought rescue in the concept of "primordial chaos." The well-ordered, "symmetrical" universe is here seen as having quickly developed from a "chaotic" and thus indefinable initial state — resulting in a final state that is relatively independent of the conditions at the start. The chaotic cosmology based on this hypothesis has not yet proved able to supply the desired explanation for cosmic symmetry, and the field can be said for the moment to belong to the anthropic principle: the initial state of the cosmos had to be as regular as it was in order that life could arise; a less organized Big Bang would have prevented the existence of intelligent observers.

If at this point it seems that modern cosmology is attempting to pull itself up "by its own bootstraps," it is probably appropriate to consider the "Bootstrap Principle" next. The Bootstrap Principle is a further development of the principle put forward by Ernst Mach, which in turn goes back to Berkeley. According to Mach's Principle, the inertia of a body is determined by the distribution of matter (galaxies, etc.) in the cosmos. The Bootstrap Principle — suggested by the U.S. physicist Geoffrey F. Chew — enlarges this hypothesis: "The properties of any one thing are not arbitrary but, on the contrary, are bootstrapped to the properties of all things." (Harrison) Thus, just as we speak of pulling oneself up by the bootstraps, or as Baron Munchhausen of legend once yanked himself from a swamp by his own hair, the hypothesis expresses the idea that no effect exerted by any agent is independent — it will result in an effect eventually being felt by the agent itself.

This curious thesis can be illustrated by an example from particle physics. If the collisions between particles involve sufficiently high energies, the end products may be more complex than those which are input. And if, as the Unified Theory predicts, quarks can change into leptons and vice versa, conversion chains are possible between all existing particles. In this way the entire world of particles would be interlinked, just as in a more general way the Bootstrap Principle demands.

232

Broadening the application to the whole of nature and taking it to its extreme, Chew arrives at the statement that "nature is as it is because this is the only possible nature consistent with itself." Harrison considers the Bootstrap Principle in its general form to be a "breathtaking concept that embodies and transcends Mach's rudimentary, mechanistic idea." But he also comments: "The basic quantities of the physical world, such as the constants of nature, seem at present to be accidental and inexplicable. But in the bootstrap picture the universe is a self-consistent whole and therefore can contain nothing of a fundamental nature that is purely accidental. According to the bootstrap picture the universe is what it is because it is consistent with itself, and it follows therefore that we are not free to separate out accidental properties and distribute them with various values among different universes, as in the anthropic picture.

"If the bootstrap principle were formulated scientifically, it would revolutionize cosmology. Until that happens, if ever, it is interesting to note that the anthropic principle serves as a makeshift or poor-man's bootstrap. It relates organisms and the universe, and although it is only a partial bootstrap, it nonetheless is useful until better ideas are found."

The bootstrap principle and the anthropic principle provide differing attempts at explaining nature. To give a further example of the first of these: Why are all electrons physically identical? The bootstrap principle would see this fact as an expression (to be formulated accordingly) of the qualities of the entire universe. The answer given by the anthropic principle is specific: only elementary particles identical in their physical characteristics and having half-value spin — like the electron — are effected by the Pauli exclusion, which in the last instance is responsible for organizing electrons in shells within the atom. And it is this very effect that takes care of the particular chemical properties of the elements which are essential to both inanimate and animate matter.

CAN WE BUILD A SIMPLER UNIVERSE?

Someone who has followed the considerations of the anthropic principle with curiosity could be tempted to outline an alternative universe which brings forth intelligent observers in a simpler manner. Could one sketch a kind of cosmic life plan in which the same goals are reached in the framework of a universe which manages with fewer "ingredients"? The universe surrounding us is already composed of remarkably few elementary particles and fields. These are the four particles of the "first generation" (see Chapter III) — proton, neutron, electron, and neutrino — and the four fundamental forces — gravitation, the electromagnetic interaction, the weak and strong nuclear force.

The simplest idea would certainly be to dispense with several of the forces and particles. Let us imagine we retained only gravitation and neutrons. An atom, in which two neutrons would be bound together only by gravity would have an extension of around one million light-years. This may be remedied by increasing the strength of gravity. If gravity were 10^{40} times stronger, and with it the electromagnetic force were increased comparably, then the atoms made from two neutrons would be about as big as normal atoms. That would have another, more drastic result. Stars would immediately become unstable, collapse, and through such vehement gravity in a moment transform themselves into black holes. Spontaneously, small pieces of matter would suffer the fate of gravitational collapse.

One could also decide — as the second possibility for simplification — that there are only two large categories of elementary particles, namely the quarks (as components of all hadrons) and the leptons. In the ideal case one could wish for only a single family of particles, bound by only a single force. This goal is strived for in the step-by-step synthesis of various theories. Indeed, within the unified theory only gravitation is now not included — perhaps it is fundamentally different — but the other three forces in the framework of the synthesis are becoming more and more similar. "Weak, strong, and electromagnetic forces could all be described in the same way with the theories," the particle physicist Howard Georgi observed not too long ago. In this way nature becomes simpler in a certain sense — not through omission of individual interactions, but rather through the recognition that the forces function similarly. "The three forces remain differentiated, but one could recognize that they operate by the same mechanism. The new theory combines leptons and quarks into one family (the "generations") and sees to it that every kind of particle can be converted into another. At the same time, the weak, strong, and electromagnetic forces are understood as aspects of a single underlying force" (Georgi). The point of this is that such completely different kinds of forces and particles despite their differences come under one overarching theory into relation with each other. "The unified theory does not attempt to conceal the differences, but it asserts they are not fundamental. The differences are conspicuous mainly because the universe is now quite cold, so that particles have low energy. If experiments could be done at extremely high energies, the unification would become apparent in all its simplicity. Leptons and quarks would be freely interconverted and the three forces would all have the same strength."

The rapid cooling of the universe accentuated drastic differences between the forces at low energy. Only through this cooling could quark and lepton structures be formed in an initially homogenous mush of cosmic rays, *only in a cold cosmos could evolution come into motion.*

An essential course of evolution is that at every new and higher level of living systems new forms of organization appear (see Chapter VIII). These are made possible through trial and error at every level of development; in each case, out of a multiplicity of varieties, the most well-adapted optimally multiply. A star which creates constant conditions for billions of years is an indispensable assumption.

Although evolution certainly strives toward no predetermined goal, it does move itself de facto in the clear direction of higher complexity and greater structural variety. An element of "freedom" is therefore constantly present. "The formation of new substances from a mixture of chemical compounds is not always the same, but varies with the temperature and the density of the medium, the catalytic agents present and the energy sources available, which naturally are different from one case to another. Thus we succeed in having a multitude of solutions to a seemingly uniquely determined problem. Another example is the small perturbations which occur occasionally in the genetic code (DNA) of the genes and are responsible for the appearance of mutations in the different species. These changes in the genetic code can be produced by many different factors (radioactivity, heat, etc.) thus providing an almost infinite variety of results. Mutations continuously enrich the biological kingdom with new species, and have been the "primary force in the evolution of life on Earth," Papagiannis writes.

The interplay of exactly four (or five) forces appears therefore to have been a very economical means of setting evolution in motion through variation and selection. The outer conditions in the planetary system are in the same way the product of a subtle interplay between these four forces. This "unity of nature," in which almost every local condition (on the Earth, for instance) is intimately tied to the entire cosmic happenstance and is dependent on it, may be unparalleled. Not only the natural laws of the structure of the cosmos, but also the special course of evolution worked together within the network of relative force relationships in an almost unique way in order to bring about an intelligent civilization. Whether there could be other, equally harmonious values for the fundamental constants eludes our knowledge. Indeed just the subtle mixture of simplicity and complexity, of cosmic, anthropic accidents and evolutionary inevitabilities makes it rather unlikely. But because we don't yet exactly know all the prerequisites for terrestrial life, this question must naturally remain open.

So nature, whose most integral component is Man, appears as the most simple and perhaps the only possible nature which can develop from the Big Bang to intelligent life. Certainly, it was not the goal (or indeed the point) of this universe to bring about observers which can "recognize" it. We can even ask ourselves, what role does Man think to play in this cosmos? It is conceivable that in the future Man will tap into more powerful sources of energy and will possess stronger natural forces, and that he will move away from the passive position of observer in order to take an active part in cosmic events. Nevertheless, he can just as quickly catapult himself out of this universe. It is in the "freedom" of human evolution that Man shows promise for both possibilities, and it is uncertain which path he will take. As Freeman Dyson once said, "it would not be surprising if it should turn out that the origin and the destiny of the energy of the universe cannot be completely understood in isolation from the phenomenon of life and consciousness."

In his "Theodice," Gottfried Wilhelm Leibniz praised this world as the best of all possible worlds and God as its creator. If the world moreover is supposed to be the only one possible, then it is simultaneously also the worst of those possible. Whether the world is the work of blind chance or the product of divine creation, whether it is the worst or the best of all conceivable alternative worlds: in the most probable singular world with intelligent observers it will still be the task in the future for us to acknowledge the inseparable connection between life and the physical world in the sense of the anthropic principle and for us — as acknowledgers of the cosmos — to try to maintain that life as well.

Appendix

Proton	Formula	
proton radius	$\hbar/m_p c$	$\sim 10^{-13}$ centimeters
proton time	$\hbar/m_p c^2$	$\sim 10^{-23}$ seconds
Electron		
electron radius (classic)	$e^2/m_e c^2$	$\sim 10^{-13}$ centimeters
electron radius (Compton wave length)	$\hbar/m_e c$	$\sim 10^{-10}$ centimeters
electron time (classic)	$e^2/m_e c^3$	$\sim 10^{-23}$ seconds
electron time (Compton time)	$\hbar/m_e c^2$	$\sim 10^{-20}$ seconds
Pion		
pion radius	$\hbar/m_\pi c$	$\sim 10^{-13}$ centimeters
pion time	$\hbar/m_\pi c^2$	$\sim 10^{-23}$ seconds

Elementary Lengths and Times

energy of the photon $\quad E \quad = \quad$ (Planck's Quantum Effect)

$\qquad\qquad\qquad\qquad\qquad x$ (frequency ζ) $= \hbar\zeta$

neutrino mass $\qquad m_v \quad = \quad (5 \pm 4)$ electron volts$/c^2$

$\qquad\qquad\qquad\qquad\qquad$ (uncertain as of 1981)

electron mass $\qquad m_e \quad = \quad 9.1 \times 10^{-28}$ grams

proton mass $\qquad m_p \quad = \quad 1.6725 \times 10^{-24}$ grams

neutron mass $\qquad m_n \quad = \quad 1.6748 \times 10^{-24}$ grams

pion mass $\qquad m_\pi \quad = \quad 0.238 \times 10^{-24}$ grams

Relationships between masses

(electron mass) / (proton mass) = 1/1 837

(pion mass) / (proton mass) = 1/7

$$\frac{\text{(neutron mass)}-\text{(proton mass)}}{\text{(neutron mass)}} = \frac{1}{730}$$

The Masses of Several Elementary Particles

Hubble constant
(determined by Sandage, 1971)

$H_o = 1.6 \times 10^{-18} \text{sec}^{-1}$
$= 50$ km/(sec megaparsec)

where
1 megaparsec $= 3 \times 10^{24}$ cm

age of the universe ($\approx 1/H_o$)

$T_o = 2 \times 10^{10}$ years
$= 0.6 \times 10^{18}$ sec

mean density of matter
in the universe

$\rho = 10^{-30}$ g/cm^3

number of baryons in the universe

$N = 10^{80}$

"length" of the universe

$R_o = cT_o \approx 10^{28}$ cm

"mass" of the universe

$M \approx \rho R_o^3 \approx 10^{53}$ g

Elementary Particle Masses and the Cosmos
pion mass

$m_\pi \sim (\hbar^2 H_o/Gc)^{1/3} \sim 10^{-24}$ g

electron mass

$m_e \sim (e^4 H_o/Gc^3)^{1/3} 10^{-27}$ g

elementary particle mass =
$\dfrac{\text{(Planck-Mass)}}{\text{(number of particles in the universe)}}$

$= \dfrac{10^{-4}\text{g}}{(10^{80})^{1/4}} \approx 10^{-24}$ g

Cosmic Fundamental Parameters and Several Cosmic-Microphysical Relationships

240

speed of light	$c = 3 \times 10^{10}$ cm/sec
Planck's quantum effect	$\hbar = 1.05 \times 10^{-27}$ erg sec
electron charge	$e = 4.8 \times 10^{-10}$ (erg cm)$^{1/2}$
gravitational constant	$G = 6.7 \times 10^{-8}$ erg cm g$^{-1/2}$
constant of the weak interaction	$g_f = 1.4 \times 10^{-49}$ erg cm^3
constant of the strong interaction	$f = 3.9$
Bohr radius of the hydrogen atom	$a_o = 10^{-8}$ cm
Planck mass	$m_p = (\hbar c/G)^{1/2} = 2.2 \times 10^{-5}$ g
Planck time	$t_p = (\hbar G/c^5)^{1/2} = 5.3 \times 10^{-44}$ sec
Planck length	$L_p = (\hbar G/c^3)^{1/2} = 1.6 \times 10^{-33}$ cm

Fundamental Constants of Nature

gravitation fine structure constant ("Alpha–G")	$\alpha_g = \left(\frac{m_p^2}{\hbar c}\right) \cdot G = 0.5 \times 10^{-40}$
fine structure constant of the weak interaction ("Alpha–W")	$\alpha_w = \left(\frac{m_e^2 c}{\hbar^3}\right) \cdot g_f = 10^{-11}$
electromagnetic fine structure constant ("Alpha–E")	$\alpha_e = \left(\frac{1}{\hbar c}\right)e^2 = 0.0073$
fine structure constant of the strong interaction ("Alpha–S")	$\alpha_f = f = 3.9$

Physical Fine Structure Constants

Group	Examples	Internal Structure	Rest mass (MeV)	Spin
photons	photon	—	0	1
leptons	neutrino	none	>0.000001	
	electron	none	0.511	1/2
	muon	none	105.66	1/2
	tau	none	1785	
mesons	pions	quarks	135	0
	kaons	quarks	497.7	0
baryons	proton	quarks	938.3	1/2
	neutrons	quarks	939.6	1/2
	hyperons	quarks	1200-1700	1/2

The Most Important Elementary Particles

	electron	muon	tau
mass energy (MeV)	0.51	105.7	1785
lifespan	stable	2.2×10^{-6} sec	$< 3 \times 10^{-12}$ sec
accompanying neutrino	electron neutrino	muon neutrino	tau-neutrino
mass of the accompanying neutrino	several eV (?)	<0.57 MeV	< 250 MeV
ratio of lepton mass to electron mass	1	207	3510
generation	I	II	III
discovered	1890s	1930s	1978

A Family of the Elementary Building Blocks of Matter: the Leptons

Along with quarks, leptons belong to the particles which cannot be further broken down into smaller subparticles. (Whether ideas about subquarks are confirmed must remain for the future.) In any case, no experiment in which leptons have been bombarded with other high-energy particles has until this time yielded any indication of a lepton "inner structure."

Most important quality of leptons: the strong nuclear force has no effect on them. Leptons react to electromagnetic, weak and naturally to gravitational forces. What unites all leptons —their electric charge is the same although their mass varies greatly.

hadron	quark structure	mass (GeV)	spin	lifespan (sec)	electric charge e
proton	uud	0.938	1/2	stable	+1
neutron	udd	0.940	1/2	10^3	0
lambda	uds	1.116	1/2	10^{-10}	0
charged lambda	udc	2.260	1/2	?	0
pi-plus	$u\bar{d}$	0.140	0	10^{-8}	+1
pi-zero	$n\bar{u} + d\bar{d}$	0.135	0	10^{-16}	0
pi-minus	$d\bar{u}$	0.140	0	10^{-8}	−1
K-plus	$u\bar{s}$	0.494	0	10^{-8}	+1
K-minus	$s\bar{u}$	0.494	0	10^{-8}	−1
phi	$s\bar{s}$	1.020	1	10^{-22}	0
psi family	$c\bar{c}$	3.1 –3.7	1	10^{-20}	0
d-zero	$c\bar{u}$	1.863	0	?	0
d-plus	$c\bar{d}$	1.863	0	?	+1
f-plus	$c\bar{s}$	?	0	?	+1
upsilon family	$b\bar{b}$	9.4 -10.5	0	10^{-8} - 10^{-11}	0
B-meson	b + (?)	5.2			0

The heaviest particles in nature, the hadrons are divided into two groups, baryons and mesons. All hadrons are formed by three quarks, mesons by two. Only five of the six possible quarks appear: up, down, strangeness, charm, and bottom. One quark and one anti-quark–indicated by a dash–make up a meson. The first hadron which implied the existence of "charm" was the psi-meson, whose "charm" however is "hidden" because the total charm of charm plus anti-charm equals zero. In contrast, the d- and f-mesons carry "naked" charm.

Bibliography

The bibliographic entries are ordered alphabetically under the Chapter in which they are utilized. Quotations in the text are usually referenced by name, allowing, through the alphabetical ordering in each chapter, easy identification of the respective sources. Sources that treat cross-referenced topics are quoted in several places.

CHAPTER I

Breuer, R.   *Contact with the Stars*, Freeman & Co., San Francisco 1982.

Barrow, J.D., Tipler, F. J.   *The Anthropic Principle*, Oxford University Press, Oxford 1986.

Carr, B.J., Rees, M. J.   "The Anthropic Principle and the structure of the physical world", *Nature* 278, p. 605 (1979).

Carter, B.   "Large Number Coincidences and the Anthropic Principle in Cosmology", in: M.S. Longair (ed .), *Confrontation of Cosmological Theories with Observational Data*, IAU-Symposium, p. 291 (1974).

Carter, B.   "The Anthropic Principle and its implications for biological evolution", *Philosophical Transactions of the Royal Society London A*, vol. 310, p. 347.

Davies, P. C. W.   *The Accidental Universe*, Cambridge Univ. Press, Cambridge 1982 .

Davies, P. C. W.   *God and the New Physics*, J. M. Dent & Sons Ltd, London 1983.

Eigen, M.   "Selforganization of Matter and the Evolution of Biological Macromolecules", *Die Naturwissenschaften 33a*, p. 465 (1971).

| Harrison, E. R. | "Cosmological Principles II. Physical Principles", *Comments in Astrophysics and Space Science 6*, p. 29 (1974). |

Rees, M., Ruffini, R., Wheeler, J. A.   *Black Holes, Gravitational Waves and Cosmology*, Gordon & Breach Sci. Publ. (New York, London, Paris, 1974), chap. 18.

Sciama, D.W.   *The Unity of the Universe*, Doubleday, Garden City 1969.

CHAPTER II

Bondi, H.   *Cosmology*, Cambridge University Press, London 1961.

Dicke, R. H.   "Dirac's Cosmology and Mach's Principle", *Nature Letters 192*, p. 440 (1961).

Dirac, P.A.M.   "Cosmological models and the Large Number Hypothesis", *Proceedings of the Royal Academy A 338*, p. 439 (1974): Dirac's first article on this topic appeared *ibid. A 165*, p. 199 (1938).

Kreuzer, J., Ellis, G.F.R.   "Physical Laws and the existence of intelligent beings", *unpublished*, Universität Kapstadt (1979).

Lévy-Leblond, J.-M.   "On the Conceptual Nature of the Physical Constants", *Rivista del Nuovo Cimento 7*, p. 187 (1977).

Harrison, E. R.   "The cosmic numbers", *Physics Today*, December, p.30 (1972).

Whittaker, E.   *From Euclid to Eddington*, A study of Conceptions of the External World, AMS Press, New York (repr. from 1949 ed.)

CHAPTER III

Bekenstein, J.

"Astronomical Consequences and Tests of Relativistic Theories of Variable Rest Masses", *Comments on Astrophysics 8*, p. 89 (1979).

Chargaff, E.

Unbegreifliches Geheimnis, Klett/Cotta, Stuttgart 1980.

Close, F.

"Particles play the generation game", *New Scientist*, November 29, p. 70 (1979).

Crease, Robert P., Mann, Charles, C.

"The Second Creation", MacMillan Publ. Co., New York 1986.

Dodd, J.

"Colouring the quark theory", *New Scientist*, March 1, p. 664 (1979).

Dodd, J.

"Universal supersymmetry", *New Scientist*, August 23, p. 597 (1979).

Drell, S. D.

"When is a particle?" *Physics Today*, June, p. 23 (1978).

Dyson, F. J.

"Ground-State energy of a finite system of charged particles", *Journal of Mathematical Physics 8*, p. 1538 (1967).

Espagnat, B.

"The Quantum Theory and Reality", *Scientific Amer-, ican* November, p. 128 (1979).

Freedman, D. Z., Nieuwenheuzen, P. V.

"Supergravity and the Unification of the Laws of Physics", *Scientific American*, February, p. 126 (1978).

Fritzsch, H.

"Flavourdynamics of Quarks and Leptons", *Physics Reports* (1981); *Quarks – Urstoff unserer Welt*, Piper Verlag, Munich - Vienna 1981.

Georgi, H.

"A Unified Theory of Elementary Particles and Forces", *Scientific American*, April, p. 40 (1981).

Georgi, H., Glashow, S.L.

"Unified theory of elementary particle forces", *Physics Today*, September, p. 30 (1980).

Glashow, S. L.

"Grand Unification: tomorrow's physics", *New Sci-, entist* September 18, p. 869 (1980).

Heisenberg, W.

"The nature of elementary particles", *Physics Today 29*, p. 32 (1976).

Heisenberg, W.

"Schritte über Grenzen", Piper-Verlag, Munich-Zurich, 1971.

Hooft, G.

"Gauge Theories of the Forces between Elementary Particles", *Scientific American 242*, No. 6, p. 104 (1980).

Jeans, J.H.

The Mathematical Theory of Electricity and Magnetism, Cambridge University Press, Cambridge 1915, 3rd ed. p. 168.

Lawrence, J. K.,
Szamosi, G.

"Statistical physics, particle masses and the cosmological coincidences", *Nature 252*, p. 538 (1974).

Lederman, L. M.

"The Ypsilon Particle", *Scientific American*, p.60 (1978).

Lieb, E.H.

"The stability of matter", *Reviews of Modern Physics 48*, p. 553 (1976).

Ludwig, G.   *Einführung in die Grundlagen der Theoretischen Physik*, Bd. 1, *Raum Zeit, Mechanik*, Bertelsmann Universitätsverlag, Düsseldorf 1974.

Malvey, J.   "The new frontier of particle physics", *Nature 278*, p.403

Perl, M. L., Kirk, W.T. "Heavy Leptons", *Scientific American*, April, p.50 (1978).

Perl, M. L.   "Leptons - what are they?", *New Scientist*, February 22, p. 564.

Rein, D. W.   "Neutrinos und charmante Quarks", *Umschau aus Wissenschaft und Technik*, 76, p. 669 (1976).

Reines, F., Sobel, H.W. "Test of the Pauli Exclusion Principle for Atomic Electrons", *Physical Review Letters 32*, p. 954 (1974).

Robinson, A.L.   "Cornell Evidence for Fifth Quark", *Science 209*, p. 1105 (1980); see also *New Scientist*, September 11, p. 776 (1980); *Physical Review Letters 45*, p. 219 and 222 (1980).

Salam, A.   "Gauge Unification of Fundamental Forces", *Reviews of Modern Physics 52*, p. 525 (1980).

Saller, H.   "Ertrinkt die Elementarteilchenphysik in den Quarks?", *Süddeutsche Zeitung*, August 17 (1978).

Schwitters, R. F. "Fundamental Particles with Charm", *Scientific American*, November, p. 56 (1977).

Sutton, C.   "The glue in the atom", *New Scientist*, September 11, p. 786 (1980).

Thirring, W.

"Erfolge und Mierfolge in der mathematischen Physik", *Physikalische Blätter 33*, p. 542 (1977).

Waloschek, P.

"Die Jagd nach dem Ypsilon", *Die Zeit*, May 19, p. 62 (1978).

Weinberg, S.

Gravitation and Cosmology, J. Wiley & Sons, New York 1972, p. 619–620.

Weizsäcker, C. F. v.

Die Einheit der Natur, Wissenschaftliche Reihe No. 4155, dtv. Munich 1974, p. 207.

Weisskopf, V. F.

"Schranken der Wissenschaft", in: *Rückblick in die Zukunft*, Severin & Siedler, Berlin 1981.

Wheeler, J. A.

"Frontiers of Time", in: N. Toraldo die Franca, B. van Fraassen (ed.), *Problems in the Foundation of Physics*, North-Holland, Amsterdam 1979.

CHAPTER IV

Abell, G. O.

"Cosmology –The Origin and Evolution of the Universe", *Mercury*, May/June, p. 45 (1978).

Alpher, R.A., Gamov, G.

"A possible relation between cosmological quantities and the characteristics of elementary particles", *Proc. Nat. Ac. Sci., 61*, p. 363 (1968).

Barrow, J. D., Silk, J.

"The Structure of the Early Universe", *Scientific American*, April, p. 98 (1980).

Barrow, J. D.

"A cosmological limit on the possible variation of G", *Monthly Notices of the Royal Astronomical Society 184*, p. 677 (1978).

Bishop, N. T.
"Time-varying gravity and the abundance of helium", *Monthly Notices of the Royal Astronomical Society 188*, p. 839 (1979).

Brown, I.M.
"The idea of the neutrino", *Physics Today*, September, p. 23 (1978).

Canuto, V.M., Hsieh, S.-H., Owen, J.R.
"Varying G", *Monthly Notices of the Royal Astronomical Society 188*, p. 829 (1979).

Collins, C.B., Hawking, S.W.
"Why is the Universe Isotropic?", *The Astrophysical Journal 180*, p. 317 (1973).

Comes-Bottaro, F.
"Les recontres de galaxies", *La Recherche*, February, p. 164 (1980).

Dirac, P. A. M.
"Cosmological Models and the Large Numbers Hypothesis", *Proceedings. of the Royal Astronomical Society London A 165*, p. 199 (1938); *A 333*, p. 403 (1973).

Dyson, F. J.
"The Fundamental Constants and Their Time Variation", in: A. Salam, P. Wigner (eds), *Aspects of Quantum Theory*, Cambridge University Press, Cambridge 1972.

Eichendorf, R.M.
Reinhardt, M.
"How Constant are the Physical Quantities?", *Zeitschrift für Naturforschung 32a*, p. 532 (1977).

Ellis, G.F.R.
"The World's Environment: The Universe", *South African Journal of Science* (1981).

Ellis, G.F.R.
Harrison, E.R.
"Cosmological Principles I. Symmetry Principles", *Comments in Astronomy and Space Physics 6*, p. 23 (1974).

252

Gamov, G.   "Electricity, Gravity and Cosmology", *Physical Review Letters 19*, p. 759 (1967).

Harrison, E. R.   "The Paradox of the Dark Night Sky", *Mercury*, July/August, p. 83 (1980).

Hawking, S.W.   "The Anisotropy of the Universe at Large Times", in M.S. Longair (ed.) *Confrontation of Cosmological Theories with Observational Data*, IAU-Symposium, p. 283 (1974).

Hawking, S.W.   "A Brief History of Time: From the Big Bang to Black Holes", Bantam Books, New York 1988.

Heller, M., Reinhardt, M.   "Meaningless Questions in Cosmology and Relativistic Astrophysics", *Zeitschrift für Naturfor-* , *schung 31a*, p. 1271 (1976).

Hönl, H.   "Kosmologische Nichtstandardmodelle und der Ursprung der Materie im Universum", *Neue Zürcher Zeitung*, May 17, p. 35 (1978).

Hoyle, F., Narlika, J. V.   "On the Nature of Mass", *Nature 233*, p. 41 (1971).

Jantsch, E.   *Die Selbstorganisation des Universums –Vom Urknall zum menschlichen Geist*, Hanser Verlag, Munich - Vienna 1979, p. 142.

Jones, B.J.T.   "The Origin of Galaxies", *Reviews of Modern Physics 48*, p. 107 (1976).

Jordan, P.   *Schwerkraft und Weltall*, Vieweg Verlag, Braunschweig 1955.

Kafka, P.	"Eintagsfliege im All", *Natur 6*, p. 80 (1981).
Kaufmann, W.	"Primordial Black Holes", *Mercury*, January/February, p. 1 (1980).
Larson, R.B.	"The Formation of Galaxies", *Mercury*, May/June, p. 53 (1979).
Lawrence, J. K.	"The Future History of the Universe", *Mercury*, November/December, p. 132 (1978).
Lindley, D.	"Primordial black holes and the deuterium abundance", *Monthly Notices of the Royal Astronomical Society 193*, p. 593 (1980).
Meier, D.L., Sunyaev, R.A.	"Primeveal Galaxies", *Scientific American*, November, p. 106 (1979).
Rees, M.J.	"Cosmological Significance of e^2/Gm^2 and Related 'Large Numbers' ", *Comments on Astrophysics 4*, p. 179 (1972).
Rees, M.J.	"The Collapse of the Universe", *The Observatory 89*, p. 193 (1969).
Reinhardt, M.	"Mach's Principle – A Critical Review", *Zeitschrift für Naturforschung 28a*, p. 529 (1979).
Rujula, A.D., Glashow, S.L.	"Neutrino weight watching", *Nature 286*, p. 755 (1980).
Schmid-Burgk, J., Scholz, M.	"Der Urknall", *Umschau aus Wissenschaft und Technik 9*, p. 276 (1979).
Silk, J.	*The Big Bang*, W.H. Freeman & Co., San Francisco 1980, p. 66-67.

Symbalisty, E.M.D., Yang, J., Schramm, D.N.	"Neutrinos and the Age of the Universe", *Nature 288*, p. 143 (1980).
Turner, M.S. Schramm, D.N.	"The Origin of Baryons in the Universe", *Nature 279*, p. 303 (1979).
Turner, M.S.	"Neutrinos: The Ultimate Astrophysical Probe", *Mercury*, January/February, p. 9 (1978).
o. Verf.	"Whatever happened to neutrino mass?", *Nature 287*, p. 481 (1980).
Van Flandern, T.	"A Determination of the Rate of Change of G", *Monthly Notices of the Royal Astronomical Society 170*, p. 333 (1975).
Valle, K., Stabell, R., Wesson, P.	"Olber's Paradox", *Astrophysical Journal*, June 15, 1987.
Weinberg, S.	*The First Three Minutes*, A Modern View of the Origin of the Universe, Basic Books, New York 1976.
Wesson, P.S.	"Does gravity change with time?", *Physics Today*, July, p. 32 (1980).

CHAPTER V

Arp, H.	"Quasars, Redshifts and Controversies", *Interstellar Media*, Berkeley 1987.
Glashow, S.L., Nanopoulos, D.W.	An Estimate of the Fine Structure Constant", *Nature 281*, p. 464 (1979).

Gribbin, J.

"Link between quasars with different red shifts gains weight", *New Scientist*, October 16, p. 157 (1980).

Harrison, E.R.

"Cosmology – The Science of the Universe", Cambridge University Press, Cambridge 1981.

Lawrence, J. K.

"Gravitational Lenses and the Double Quasar", *Mercury*, May/June, p. 66 (1980).

Morrison, N.D.
Morrison, D.

"Are the Laws of Physics the Same Everywhere?", *Mercury*, July/August, p. 99 (1980).

Rees, M.J.

Galactic nuclei and quasars: supermassive black holes?", *New Scientist*, October 19, p. 88 (1978).

Refsdal, S.

Monthly Notices of the Royal Astronomical Society, Vol. 128, p. 295 (1964).

Savedoff, M.P.

"Physical Constants in Extra-Galactic Nebulae", *Nature 178*, p. 688 (1956).

Smith, H.E.

"Quasi-Stellar Objects", *Mercury*, March/April, p. 27 (1978).

Teller, E.

"On the Change of Physical Constants", *Physical Review 73*, p. 801 (1948).

Tubbs, A.D., Wolfe, A.M.

"Evidence for Large-Scale Uniformity of Physical Laws", *Astrophysical Journal Letters 236*, p. 105 (1980).

Wolfe, A.M., Brown, R.L., Roberts, M.S.

"Limits on the Variation of Fundamental Atomic Quantities over Cosmic Time Scales", *Physical Review Letters 37*, p. 179 (1976).

CHAPTER VI

Arnett, W.D.   "The Physics of Gravitational Collapse", Texas Symposium on Relativistic Astrophysics, *Annals of the New York Academy of Science*, p. 366 (1979).

Brecher, K., Lieber, E., Lieber, A.E.   "A Near-Eastern sighting of the supernova explosion of 1054", *Nature 273*, p. 728 (1978).

Fonkal, P.   "Does the Sun's Luminosity Vary?" *Sky & Telescope 2*, p. 111 (1980).

Gamov, G.   "Electricity, Gravity and Cosmology", *Physical Review Letters 19*, p. 759 (1967).

Herbst, W., Assousa, G.E.   "Supernovas and Star Formation", *Scientific American*, August, p. 122 (1979).

Hillebrandt, W.   "Sternenentwicklung und Elementsynthese I & II, *Physikalische Blätter 35*, p. 13, p. 65 (1979).

Kippenhahn, R.   *One Hundred Billion Suns: The Birth, Life, and Death of the Stars*, Basic Books, New York 1983.

Mansfield, V.N.   "Pulsar spin down and cosmologies with varying gravity", *Nature 261*, p. 560 (1976).

Morrison, N.D., Morrison, D.   "Supernova-induced Formation of Stars and Planetary Systems", *Mercury*, March/April, p. 40 (1978).

Murphy, C.T., Dicke, R.H.   *Proceedings of the American Philosophical Society 108*, p. 224 (1964).

Pochoda, P., Schwartzschild, M.   "Variation of the Gravitational Constant and the Evolution of the Sun", *Astrophysical Journal 139*, p. 587 (1964).

| Schramm, D.N. | "Neutrino Astronomy", Texas Symposium on Relativistic Astrophysics", *Annals of the N.Y. Academy of Sciences*, p. 380 (1979). |

CHAPTER VII

| Canuto, V.M. | "The Earth's radius and the G variation", *Nature 290*, p. 739 (1981). |

| Cark, D., Hunt, G., McCrea, W. | "Celestial chaos and terrestrial catastrophes", *New Scientist*, December 14, pp. 861–863 (1978). |

| Dicke, R.H. | "Principle of Equivalence and the Weak Interactions", *Reviews of Modern Physics 29*, p. 355-362 (1957). |

| Gribbin, J. (Ed.) | *Climatic Change*, Cambridge University Press, Cambridge 1978. |

| Hart, M. | "The Evolution of the Atmosphere on Earth", *Icarus 33*, p. 23 (1978). |

| Lambeck, K. | "The Earth's Variable Rotation", *New Scientist*, November 13, pp. 426–429 (1980). |

| McCrea, W.H. | "A Philosophy for Big-Bang Cosmology", *Nature 228*, p. 21 (1970). |

| O'Nions, R.K., Hamilton, P.J., Evemon, N.M. | "The Chemical Evolution of the Earth's Mantle", *Scientific American 242*, No. 5, p. 120 (1980). |

| Pollack, H.N. | "The Cooling Earth", *Nature 286*, p. 655 (1980). |

| Sprague, D., Pollack, H.N. | "Heat flow in the Mesozoic and Cenozoic", *Nature 285*, p. 393 (1980). |

| Taylor, F., Webb, G. | "Where does radiation come from?" *New Scientist*, December 21/28, p. 922 (1978). |

Tucker, W. "Supernovas, Dinosaurs and Us", *Mercury*, July/August, p. 95 (1980).

Wesson, P.S. *Cosmology and Geophysics*, A. Hilger, Bristol UK (1978). *Gravity, Particles and Astrophysics*, D. Reidel, Dordrecht 1980.

Wolfendale, A. "Cosmic rays and ancient catastrophes", *New Scientist*, August 31, p. 634 (1978).

Wood, R.M. "Is the Earth getting bigger?" *New Scientist*, February 8, p. 387 (1981).

CHAPTER VIII

Applewhite, P. B. *Molecular Gods – How Molecules Determine Our Behaviour*, Prentice-Hall, Inc., Englewood Cliffs, N.J. 1981.

Asimov, I. "Not as We Know It – The Chemistry of Life", *Cosmic Search 3* (1), p. 5 (1981).

Dose, K., Ruchfuss, H. *Chemische Evolution und der Ursprung lebender Systeme*, Wiss. Verlagsges. mbH, Stuttgart 1975.

Eigen, M. "Selforganization of Matter and the Evolution of Biological Macromolecules", *Die Naturwissenschaften 58*, p. 465 (1971).

Eigen, M., Schuster, P. *The Hypercycle*, Springer-Verlag, Heidelberg, Berlin, New York 1980.

Eigen, M., Winkler, H. Winkler-Oswatitsch, R., "The Game of Evolution", *Interdisciplinary Science Reviews I*, p. 19 (1976).

Ellis, G.R.F., Kreuzer, J.   "Physical Laws and the existence of intelligent beings" *unpublished*, Universität von Kapstadt (1979).

Gierer, A.   "Physik der biologischen Gestaltbildungs", *Die Naturwissenschaften 68*, p. 245 (1981).

Hubel, D.H.   "The Brain", *Scientific American 241*, p. 39 (1979).

Miller, J.G.   *Living Systems*, McGraw-Hill Book Co., New York 1978

Pauling, L.   *The Nature of the Chemical Bond*, Cornell University Press, 3rd Edition, Ithaca, N.Y., 1960.

Simpson, G.G.   *The Meaning of Evolution*, Yale University Press, 1976.

Slytyer, R.O.   *Plant-Water Relationships*, Academic Press, London and New York 1967.

Stevens, C.F.   "The Neutron", *Scientific American 241*, p. 49 (1979).

Tosteton, D.C.   "Lithium and Mania", *Scientific American*, April, p. 130 (1981).

Stillinger, F.H.   "Water-Revisited", *Science 209*, p. 415 (1980).

Vogel, G., Angermann, H.   *dtv-Atlas zur Biologie I & II*, dtv-Verlag, Munich 1957

Winfree, A.   "Chemical Clocks: a clue to biological rhythms", *New Scientist*, October 5, p. 10 (1978).

CHAPTER IX

Asimov, I.   *The Gods Themselves*, 1973.

Budecki, F.

"Johannes Kepler – Die Ganzheit von Mensch, Natur und Gott", *Physikalische Blätter 11*, p. 323 (1980).

Chargaff, E.

"Die Patentierbarkeit des Menschen", Transatlantik *4*, p. 36 (1981).

Chew, G.F.

"Bootstrap – A Scientific Idea?", *Science 161*, p. 762 (1968).

Dicke, R. H.
Peebles, P.J.E.

"The Big Bang Cosmology – Enigmas and Nostrums", in: *Hawking, Israel, W. (Ed.), Einstein Centenary Colume*, p. 517 (1980).

Dyson, F.J.

"Energy in the Universe", *Scientific American 9*, p. 51 (1971).

Ellis, G.F.R.
Harrison, E.R.

"Cosmological Principles I. Symmetry Principles", *Comments in Astrophysics and Space Science 6*, p. 23 (1974). "Cosmological Principles II. Physical Principles", *ibid.* p. 29 (1974).

Papagiannis, M.D.

"Could You Build A Better Universe?", *Griffith Observer*, August, p. 3 (1974).

Scholes, R., Rabkin, E.S.

Science Fiction – History, Science, Vision, Oxford University Press, London, 1977.

Stapledon, O.

Star Maker, Penguin Book Ltd., Harmondsworth 1975; Orig. Methuen, 1937.

Printed in the United States
By Bookmasters